The Pig
in Thin Air

Also in the {bio}graphies Series

Pattrice Jones
The Oxen at the Intersection
A Collision

Martin Rowe
The Polar Bear in the Zoo
A Speculation

Martin Rowe
The Elephants in the Room
An Excavation

The Pig in Thin Air

An Identification

Alex Lockwood

Lantern Books ● New York
A Division of Booklight Inc.

2016
Lantern Books
128 Second Place
Brooklyn, NY 11231
www.lanternbooks.com

Permissions

Excerpts from *Pig Tales* by Marie Darrieussecq. Copyright © 1996 by Marie Darrieussecq. Reprinted by permission of Faber and Faber (World) and P.O.L. (U.S. and Canada). Image of "Pig escapes from moving lorry, Foshan, Guangdong Province, China—01 Jun 2014" rights acquired from Rex Features.

Printed in the United States of America

Library of Congress Cataloging-in-Publication Data
Names: Lockwood, Alex, author.
Title: The pig in thin air : an identification / Alex Lockwood.
Description: New York : Lantern Books, a division of Booklight Inc. [2016] |
Includes bibliographical references.
Identifiers: LCCN 2015030949 | ISBN 9781590565353 (pbk. : alk. paper) | ISBN 9781590565360 (ebook)
Subjects: LCSH: Animal rights activists—Great Britain. | Animal rights—Great Britain. | Veganism—Great Britain. | Animal rights activists. | Animal rights. | Veganism. | Human–animal relationships.
Classification: LCC HV4805.A3 L63 2016 | DDC 179/.30941—dc23
LC record available at http://lccn.loc.gov/2015030949

For Tricia Thraves
(1947–2015)
and for Liam

For the animals,
and their thanks for
all you do.
[illegible signature]
2016.

Contents

Acknowledgments

THIS BOOK WOULD not have been possible without all I have learned about animals, activism, and writing from Alicia Craven, Gene Baur, Drew Alexis, Bruce Friedrich, Susie Coston, Liz Marshall, Lorena Elke, Hannah Gregus, Susan Banks, Jo-Anne McArthur, Jane Pearce, Michelle King, Wendy Matthews, Holly McNulty, Carolyn Mullin, Ronnie Rose, Wayne Hsiung, Brian Burns, Marychris Staples, Katie Barber, Christine Gittings, Michael Anthony, Paul York, Larry Gordon, and many others.

I particularly want to thank Anita Krajnc for her welcome and for urging me to put my thoughts into writing, as well as for the activism she performs daily.

Viccy Adams, Jill Clough, Andrew Crumey, Geoff Dunster, Louise Fisk, Simon Gill, Rob Glossop, Andrew Harding, Katie Hindle, Ceri Hughes, Elliot Jones, Hannah Kirkham, Adam Strange, Dean Tyler, Annabel Tremlett, Liam Walker, and Anne Whitehead have all played their part in supporting my writing over the years, although any errors made herein are my own.

I would also like to thank Kara Davis and Martin Rowe from Lantern Books for working with me on this project and seeing it come to life. And my sincerest gratitude goes to the staff and interview panel of the Winston Churchill Memorial Trust, who sent me on my trip.

To all the nonhuman animals I've met so far on my journey into animal advocacy, and particularly to the one pig who was there when it all began, on the other side of the world—thank you.

Introduction

IN THE SUMMER of 2014 I set out on a journey to find my place in the world of animal activism. Although I'd been a vegetarian since the 1990s and a campaigner and then scholar on climate change since 2002, I was new to animal advocacy. For much of my life I'd been uncomfortable with our species' relationship with nonhuman others and the denial of the animal roots of everyday life, yet my contribution was limited to a love for cats, birds, and wildlife, and weekends in volunteer conservation. I finally became vegan in 2010, changing my habits for the benefit of nonhuman others. I shifted my scholarship to the field of Human–Animal Studies (HAS), to which *The Pig in Thin Air* makes, I hope, a useful contribution. Yet the world of direct activism eluded me. Was it the kinds of activism available that scared me off? Or was the reason something more internal, deep-set, closer to home?

With a U.K. government culling badgers and keen on hunting foxes, and global actions against dolphin slaughter and dog meat festivals, getting involved in protest had never been more opportune. Yet the graphic imagery, the blood-covered lab coats, and the barracking of high street shoppers on a Saturday afternoon left me feeling uncomfortable and exposed. I was unconvinced about the effectiveness of the noisy marches and street stalls under banners against vivisection, fur, and circuses. Such an approach didn't seem to be answering what I saw as the central question of the movement: *How do we change our relationship as a species to other species?* Such protests, as legitimate as they are, did not bring me any closer to nonhuman others; nor did they seem to bring enough people over to what the French philosopher Jean-Christophe Bailly has called "the animal side." I wanted to discover what else there was—a way to be effective as an activist that brought connection with the nonhuman world. I wanted to put my body on the front line of animal rights work but felt there must be the chance to draw that line a different way. I was seeking alternatives. So I packed my bag and headed to North America, which looked, on social media at least, to be where more inviting things were happening.

As a writer and academic curious to know what good my skill set could offer the movement, I met advocates, human and nonhuman, working to change the story of how we relate across species boundaries. It became clear that some of the most effective work was to be found advocating for those within our food-production systems. What I encountered changed my understanding of animals, both human and nonhuman. My identity as a human and my concept of what an advocate can be is now different. That is, the story I tell about myself has changed. How it changed, and why that matters, is at the heart of this book.

Lots of wonderful books have already been written that document the history, tactics, and achievements of the movement, many by activists with decades of experience. *The Pig in Thin Air*, rather, comes interwoven with my fifteen years' contribution as a campaigner and researcher on environmental issues, especially climate change. It is also shaped by my academic training in the scholarship of bodily emotions and practices. Combined with my personal journey, I hope what follows offers something for experienced and new advocates alike.

The Pig in Thin Air argues that animal advocates will achieve their goal of liberating nonhuman others when they aim to change not only the "rational" thoughts people have about nonhuman others, but also the embodied knowledges of who we feel ourselves to be. That is: *How we feel about who we are* dictates to a great extent *what we do (to others).*

To do this, activists must pay closer attention to the role played by bodies in bringing about change in behavior for nonhuman benefit. This is not as easy as it sounds. How selves are formed through embodied knowledge—how we feel and act, which shapes how we think—has been, in Western histories, marginalized in favor of the rational and transcendent. But developments in philosophy, the social sciences, and psychology are challenging traditional concepts of the "body," "self," and "identity"—and therefore how they might change. *The Pig in Thin Air* explores this dynamic from the body up, to reanimate the communities of advocates working to end the suffering of nonhuman animals, before our time runs out. Because our time, in terms of the threat of climate change, *is* definitely running out. An end to animal suffering could be the very thing that saves us all.

—

I was supported on my trip by a Winston Churchill Travel Fellowship. The fellowship provides an opportunity for citizens of the United Kingdom to visit other countries, learn from their ways of doing things, and use that knowledge to benefit the relevant communities of practice. The primary mission of the fellowship is to help build world peace. Its vision is that if we immerse ourselves in each others' cultures, then there is less desire to go to war. Such a vision is relevant to our warring relationship with nonhuman animals. At every turn of my fellowship I discovered something new about the nature of embodied knowledge. I learned of the bond that emerges when humans meet nonhumans as their advocates. I came to see that not only are we advocates for them, but in the true sense of the word—as a call to aid—our nonhuman companions, in their embodied wisdom of the crises we face, are also advocating for *us*.

At the heart of *The Pig in Thin Air* is the argument that when we meet, live with, and mourn, body-to-body, the lives and deaths of individual nonhumans, it changes us via the embodied knowledge of actively witnessing the experience. When we *do* spend more time with those individuals, as we learn experientially about their ways of life and, critically, do this with what the philosopher Lori Gruen has called "entangled empathy" (a form of moral attention), then we can and do choose to stop being at war with them, too.

Is the metaphor of war a suitable one? Is it a metaphor at all? As many have argued, from philosopher Jacques Derrida to photojournalist Jo-Anne McArthur to the contributors in the collection *Terrorists or Freedom Fighters? Reflections on the Liberation of Animals*, humans *are* waging a global all-out war on the nonhuman animal. This is most marked in our consumption of their bodies for food and other consumer products, but it is also enacted in the thousand other disposable uses to which we put nonhuman bodies to work without their consent. Jo-Anne McArthur identifies herself as a war journalist. I have come to see the path open to me as a war reporter, although my work is far from *Death in the Afternoon*. As a book interested in the stories we tell ourselves about ourselves, *The Pig in Thin Air* aims to illustrate what value narrative has for changing our relationship with nonhuman others. We need to have a different story to tell about our species if a shift is to be both global and enduring.

Such change cannot be enacted through story alone. Nor can it be done solely through the circulation of imagery and information via social media, as useful as these networks have proven to be. Stories and imagery must be put to work at the corporeal level beyond language. Or as philosopher Mary Midgley phrases it, "What makes our fellow beings entitled to basic consideration is surely not intellectual capacity but emotional fellowship" (60). We engage in this "emotional fellowship" by recognizing the absolute value of bodily freedom and our obligations toward ensuring that all animals who want to express their natural desires can. The lives of animals, to borrow a phrase from J. M. Coetzee, will not be known to us best by consideration through rational thought, but when we come to feel their desires in our bodies.

If we do so, we may just protect the air we all breathe and that sustains this vulnerable planet. At the moment our bodies are in suspension. When it comes to climate change, it seems we have yet to reconcile the facts of the worsening situation with our wish not to believe them. *The Pig in Thin Air* trades in this suspense—*what will happen?* But it also deals with another kind of suspension, from the Latin *suspendere*: "to hang up; kill by hanging." It is the body of the pig known to us in this vulnerable position—held up by the back leg on the kill line—that offers a stark warning that we, too, have been left hanging over an abyss, suspended by the inaction on climate change by our political and corporate classes—and our own inertia, fear, or complacency. We need to find a way out of this state and move toward the issue; otherwise, we just might find there's nowhere left to go except downward. Naomi Klein calls for such a movement in her book *This Changes Everything: Capitalism vs. the Climate*. She also turns to war as a (literal) explanation of the situation: "Our economic system and our planetary system are now at war. Or, more accurately, our economy is at war with many forms of life on earth, including human life" (21). It is with our bodies that we must fight back.

—

The Pig in Thin Air is organized around two questions, following a journey of learning how I could put my body into advocacy work, and using this as an example of the way identity can and does change.

Part I asks: *What makes an animal advocate?* There are many potential answers to this. So I ask: *What made* me *an animal advocate?* My progress was instigated not only by what I thought about nonhuman others, but what I was able to feel about myself. This was made possible by a painful journey from ignorance to awareness of the destructive habits in my personal life. Chapter 1 tells the story of how I changed these habits, and how this change opened me up to do more for other bodies, particularly those tortured bodies of pigs. Or to put it another way, to become an effective advocate I first had to change how I related to my own body and the damaging actions that flowed from it. But *The Pig in Thin Air* is not about an extraordinary conversion. In fact, quite the opposite: it is from the lowly or less-than vantage point of ordinary habits that we are able to know (and feel) what we talk about when we talk about bodies. Philosophers George Lakoff and Mark Johnson write in *Philosophy in the Flesh* that what we know now from cognitive science about the embodied mind "utterly changes our relation to other animals and changes our conception of human beings as uniquely rational" (4). Yet what people are able to think and feel about themselves is often ignored in the ways we approach behavioral change. Reclaiming and strengthening our connections to embodied experience offers the potential for radical change in relation to the nonhuman.

The journey I made from ignorance to awareness of the suffering of nonhumans was accelerated, as it has been for many, through social media access to imagery. So in Chapter 2 I dive into how social media helped move me from vegetarian to vegan living, and beyond to advocacy. We meet our titular pig and I tell the story of how our bodies became entangled. Her body, the story around her body, and the air around the story are all implicated in the chance to change how we live.

In Chapter 3 I explore the link between animal advocacy, self-identity, and the crisis of climate change. Here I ask: *Does working across animal rights and climate change together offer the best hope for animal equality and justice, and a safe future for everyone?* This idea has been called by some "climate veganism," responding to the disastrous impact of animal agriculture on our environment. In this approach, vegan life practices are essential in building a sustainable ecological future. What I want to examine is whether the effectiveness of any response to both climate change and animal rights essentially depends on not only environmental and animal activists working together, but on having radically different senses of who we are. My argument is that broad cultural and behavioral change will happen if we learn to form new bodily identities able to respond to and counter the crises we face, to do so in new and productive ways. That is, *change is always bodily change.* A reframing of the self—not as autonomous, but as connected and vulnerable—is the basis of this work. How do we do this before the air we breathe gets too hot?

With this bodywork done, Part II takes us on the journey of my fellowship over the summer of 2014. Here I ask: *How can I be a good animal advocate?*

In Chapter 4 I take my first steps as an advocate and advocate-writer, exploring narratives of change across genres that tackle the systemic behaviors driving the worst abuses of nonhuman animals within the food system. These narratives help us question how a body of text and a body of work, and the corpus of animal stories across the world, bring our attention to the ways in which narrative shapes our understanding of the vulnerable corpus of the animal body; in this case, especially the pig.

In Chapter 5 we get on the road and visit America's most successful animal rescue and advocacy organization, Farm Sanctuary, to see the other side of the lives of farmed animals rescued from exploitative production, and experience the powerful bonds between humans and nonhumans shaping and healing the advocacy movement.

Then in Chapter 6 I look at the connection between human and nonhuman advocates made through bearing witness. When performed at vigils outside slaughterhouses and during silent protests in city centers—critically, when the body of the nonhuman animal is present—then active witnessing creates new embodied knowledge that brings us much closer to the goal of "emotional fellowship." But I wanted to go a little further and enact my desire to advocate for the animals in a way that drew a new line. So in Chapter 7 you'll join me on a viscerally challenging, physical action that taught me, on the final day of my trip, what exactly I was going to do with the rest of my (advocate) life.

This book does not cover all of the new and emerging forms of animal activism that place the relationship between bodies front and center. It is, rather, a snapshot of ongoing research into what makes animal advocacy work. *The Pig in Thin Air* opens a conversation about the body that is central to the future of effective animal advocacy. What we do with, put into, and exclude from our bodies are acts that produce our identities. This book is the story of these bodies finding ways to relate to each other—some surprising, others merely practical, but all, as Lori Gruen puts it, entangled.

It is easy to lose sight of the individual nonhuman. To counter this we must learn instead how to feel their suffering even when they are out of sight. And although the narrative of *The Pig in Thin Air* spreads outward from the individual that gives this book its name, it will always come back to her: her life, her suffering, her body. She was the catalyst for the questions laid out here. We shall meet her soon. But before I was able to come close to her, I needed to move nearer to another body. My own.

Part I

What Makes an Animal Advocate?

1

The Breaking Down of Bodies

OUR BODIES ARE central to how we live; they are the corporeal forms through which we experience the world and relate to others. How we mobilize, act, and respond, or are impeded from doing so, are the central questions of our lives. Such questions entail the quality of our freedom and flourishing.

Our bodies are the source of how society is structured. As the sociologist Chris Shilling puts it: "It is the properties and capacities of embodied humans that provide the corporeal basis on which identities and social relations are consolidated and changed" (256). Or to put it another way, our attractions and repulsions to environments and other people are embodied visceral experiences that provide the motivations for how we maintain, develop, or transform the world. It is never only our "disembodied" intellectual evaluations that guide action, but always thoughts entangled with bodily reactions that become knowable and nameable as emotions: joy, disgust, anger, fear.

Yet many sociologists and philosophers have left us with the feeling that bodies are barely relevant to the creation of social life. The Cartesian and Kantian traditions of rationalist ethics—emanating from the mind—have rejected, for example, compassion and sympathy—emotions arising from the body—as a basis for making moral decisions. Contemporary philosophers Peter Singer and Tom Regan have also been criticized for steering away from our emotional response to animal suffering. This predominantly male, white view of social life has focused on abstract, universalist processes and left out particular bodies within those processes. Where bodies have turned up they are almost exclusively human, usually as irrational influences on moral choice. Nonhuman exceptions prove the human rule. As Rhoda Wilkie has pointed out, those from anthropocentric cultures tend to reinforce such a worldview and separate out human and nonhuman bodies in studying how social worlds are made.

Such worldviews are being challenged. Barbara Noske has argued that the beings who make our social world work are not only human. Such beings are all individual selves—a single human, a single pig. To pay these individuals consideration is to give what Margaret Urban Walker has called an "attention to the particular" (166) grounded in the feminist ethic-of-care tradition. *The Pig in Thin Air* is an attempt to take particular bodies into account. I want to do this without generalization, and without losing sight (or sound, or smell) of the bodies involved.

We could begin with any nonhuman animal body. But, at the risk of perpetuating the anthropocentric worldview, what about my own? How is it that I've come to try and shape this world as an animal advocate? This story may be worth telling for those involved in bringing people over to "the animal side," because for two decades as a vegetarian, environmental journalist, and climate activist I never considered working for animal liberation. It was only when I began to listen to my body after years of intellectualizing myself away from my corporeal sense of who I was that I was able to act compassionately for these other bodies. To look at my body will, I hope, not reinforce a speciesist and patriarchal position, but illuminate the damaging nature of a rationalist worldview that excludes messy materiality from moral action. So the body with which this story begins is my own.

I grew up during the 1970s and 80s in various council-owned flats in a one-parent household in South London. Following my parents' divorce, my mother, sister, and I moved around often, and my mother faced financial difficulties bringing up two children as a single parent. We were working class and urban, and I didn't have any special connection with the natural world or to animals. For the first eight years of my life, our family did not have a garden. I would sometimes spend time with my grandfather in his. Later, when my mother remarried, my stepfather kept an allotment where I learned about the Colorado beetle, potatoes, and the solitude of men in their sheds.

Growing up, our routines were structured around cheap ways to feed a family. Monday to Thursday would mean chicken parts doused in sauce from a jar. Friday nights involved burgers from McDonald's, my mother too tired to cook after a long week of work. Saturday was a visit to my father's and a choice of either frozen sausages or beef burgers with chips. Saturday evening back at my mother's was steak or pork chops. Sunday was always a cooked breakfast and a roast dinner at my grandparents'. I didn't consider where any of the food on my plate came from, nor the preponderance of meat. Why should I? The people I trusted decided on my eating habits, even when this led to my sister and I becoming overweight as my mother expressed care through food—sweets and processed meals being easier to provide on a daily basis while she worked two, sometimes three, jobs. Until I was fifteen years old, my "food identity" was formed through this affection and a reliance on comfort eating, including the eating of meat. For Roland Barthes, "When he buys an item of food, consumes it, or services it, modern man does not manipulate a simple object in a purely transitive fashion; this item of food sums up and transmits a situation; it constitutes an information; it 'signifies'" (78). What we eat has always denoted class, privilege, religion, and other key markers of identity, such as masculinity. (For instance, in 2015, McDonald's advertised a new pork sandwich with the tagline "Sausageness. Baconness. Manliness.")

Then, at fifteen, I turned vegetarian. Looking back over twenty-five years later, I see I did not make this decision after connecting my food behaviors with any sense of wanting to protect the planet and other species. Despite the everydayness of eating animals, I never consciously thought of the animal body. I turned vegetarian in the desire to alter the ill feeling developing in our family, as my mother's second marriage disintegrated amid daily arguments and a lack of affection between she and my stepfather. These were never more exposed than when we sat down for dinner. Being vegetarian meant I could avoid that, or at least disrupt it. Drawing attention to what I put in my mouth was, thus, in some ways an effort to divert attention to what came out of theirs.

My attempts at giving up meat lasted just a few weeks. By sixteen I was already a child of two divorces, renewed financial insecurity, and an alcoholic father. I grew distant from my family as I looked for multiple ways out, including changing my food identity. My life was not one of unmitigated or physical trauma, but it was nonetheless full of anxiety. My father's drinking had driven away my sister from our weekly visits and would soon push me away too, before he disappeared on a bender and went missing—and remains missing, these many years later. Through all of this, I lived in a state of constant awareness of my surroundings. Even as young as six, I was a terrible sleeper, waking at 4 A.M. each morning and lying open-eyed until we were allowed to get up. It was a pattern of poor sleep I kept for the next thirty years.

Like many who have undergone difficult upbringings, I chose to ignore the feelings swirling through me. I sought protective armor in a different corpus: the world of writing and reading. I escaped into fantasies more exciting and redemptive than my own reality, filled with the conflicts I could not resolve in real life. A stressed childhood was at least, perhaps, as both Ernest Hemingway and Doris Lessing have said, good training for being a writer. I wrote creatively, then turned to journalism, and then finally became an academic. Such a route to self-protection was made through an intellectualization of unstable emotions. I was unwilling to admit the feelings in my body and so was unable to express them as emotions to parents who were the source of much of my anger. In psychological terminology, I "closed down."

This practice became a habit, and, spurred on by academic success I sought to intellectualize, analyze, and criticize at every opportunity. I chose the more easily rewarded rationalizations of the mind and, in doing so, stopped knowing how to feel—if, indeed (as a therapist once pointed out), I'd ever properly learned the practice of appraising affects into the appropriate emotions. To live is to be able to affect and be affected by our environments, including other bodies. Affects are physiological forces: the altered body–mind states that we register as heightened heart rates, lurching stomachs, or gut feelings or sweaty palms, which we then appraise and name in their social forms as emotions, depending on context. Knowing our bodily *affects* and owning their *effects* is critical to understanding how our lives are shaped by their force. To close ourselves off from responding to our affects is to misread our bodies, and this will always lead us to fail to live fully, to flourish physically, and to relate compassionately.

We cannot close ourselves off entirely, of course—our bodies are always open to affects because they come from within. But we can intellectualize away those feelings rather than feel them, and mis-respond accordingly. I did this without understanding that this had happened, because such a narrative is normal and naturalized. Western societies are structured around a hierarchical Cartesian objectivism where rationality and the mind are considered superior and preferable to the emotional and unruly body. As feminist critics have argued, this dualistic structure permeates our relationships between men/

women, culture/nature, white/black, human/nonhuman. Separation from our body is part of the successful domination of the "natural" world of which we are, corporeally and materially, always connected through our biological systems and cultural practices. This separation is essential to the work of patriarchy and dominating ideologies that "other" those who might threaten their success. As ethicist Josephine Donovan has written, "rights theory and the Kantian rationalist ethic were developed for an elite of white property-owning males" (289). This logic is at the heart of speciesism, the ideological practice—ideologies are always dynamic practices—that arbitrarily separates our species from others. The assumption of our superiority leads to exploitation.

But the body is not separate from the mind. Rather, suggests sociologist Lisa Blackman in *Immaterial Bodies*, bodies are best defined as "processes that extend into and are immersed in worlds. That is, rather than talk of bodies, we might instead talk of "brain–body–world entanglements," and where, how and whether we should attempt to draw boundaries between the human and nonhuman, self and other" (1).

Blackman suggests that if we understand our experiences as brain–body–world entanglements we come much closer to sensory reality, and a connection to this corporeal way of being embedded in relations. Too often we skip straight from mind to world, over the body. The bodily and social norms to which we have been socialized are often unhealthy or morally problematic for us as individuals. We close off our bodily responses and subsume our individuality into normative roles to maintain social contracts. We stop listening to messages of pain and illness; we fail to see ourselves as vulnerable and mortal. We deny ourselves our feelings if they fall outside of what is accepted as "normal"—for example, the miseries of dieting in search of the ideal body shape or strict gender roles. Such norms are held as ideals in societies—benefiting a small section of (usually white, male) elites who profit from such ideals—and formulate the body to be a "project," an "option," or a "regime," each, as Shilling says, "constructed on the basis of a *chronic corporeal presence*" (220) [emphasis in original].

Embodiment as entangled process leads to many kinds of behavior, what scholar Elspeth Probyn calls "bodily orientations, ingrained ways of embodying the world, which, while they may change, are truly our history forged in the flesh, taste, and memory" (291). We embody orientations such as empathy or courage (from the Latin *cor*, meaning "heart"). We also embody cruelty, apathy, or disregard. Critically, we embody such qualities as we *enact* them, in the flesh.

Experiencing life through our brain–body–world entanglements already suggests the messy nature of who we might be. Repressing this affective power limits the flourishing of experience. Viewing experience *as* entanglement opens us to the reality of individual others affecting us. If we see our bodies on a continuum with other species and entangled, we should find it harder to exploit them. That is, as Erika Cudworth has suggested, other species are "a formation of social power" (7) that alters worldviews. And a change in worldview from the abstract and impersonal to the particular and personal is at the heart of animal advocacy. We cannot work as effective advocates if we do not take our own affective bodies seriously.

The first time I made a decision in response to what I *felt* as a searing injustice committed against other species was when I was seventeen, after reading an article in the *Guardian* weekend magazine about the hunting of whales. Although I was not eating whale meat, the injustice of the slaughter must have triggered a sense of fellow-feeling. Turning vegetarian seemed like an option—one I knew could bring about a change in circumstance, if my family's angry response to my first attempts were anything to go by. Seeing the whale's slaughter was my first felt experience of entanglement with another species, of what the philosopher Cora Diamond aptly calls a feeling of being "in the same boat" as other creatures (474). This fellow-feeling makes it possible to seek out another across species lines as *company*.

To see someone else as company is not to argue for abstract rights but, says Diamond, to feel "closely connected with the idea of something like a *respect* for the animal's independent life" (475) [emphasis in original]. To respect is to look again; to look more closely. It is to recognize that the nonhuman has a life that means something to herself. To see the whale, bloodied and dragged up the side of a Japanese whaling boat, as a self whose bodily independence we should respect—but had not—was to make a compassionate leap toward this looking again. Importantly, I'd just processed a bodily affect into emotion (a feeling of injustice at the whale's murder) and acted upon it (eschewed meat). It was a lesson that, at least within my corporeal self, I wouldn't forget. This act is written on page one of my body (of) work: the story of becoming an advocate.

This second attempt at vegetarianism didn't last long either—six months. The feelings of injustice waned, lost amongst the confusion of teenagerhood and my mother falling into financial insecurity and illness caused by her second divorce and overwork. Throughout my early twenties, I continued to eat animal products. I remember my friends and I at university buying *more* steak as the price of beef dropped due to what came to be known commonly as mad cow disease. I recall being jovially concerned that I was at high risk of developing new variant Creutzfeldt-Jakob disease (nvCJD), the condition in human form, because as a child I'd eaten so much cheap processed meat, the likely source of crossover from cows to humans.

Yet in my mid-twenties I returned to being vegetarian, and (mostly) stayed that way. It wasn't to do with taste: I liked the mouthfeel of animal flesh and the role that meat and fish played in my identity. At the time, I knew nothing about veganism or animal liberation. I'd never knowingly met a vegan. Somehow, the body of that whale lingered in my viscera. I remembered the positive act of feeling my feelings. An entanglement that had shown itself across continents, stimulated by the image, left its trace. I understood that to eat the body of others was to ignore the fellow-feeling I'd kept for the whale in my body.

I didn't go further. I didn't give consideration to animals other than those I thought under threat, such as whales or domestic companions. They had my fellow-feeling. I obviously *felt* vegetarianism to be enough. That is, I considered this specific act of feeling a far enough excursion into an affective life I'd always suppressed.

It was only in 2010, at the age of thirty-five, that I awoke to the horrors of the way that we exploit, abuse, laugh at, prod, poke, and subordinate animals. Needless to say, it was a painful learning process. As well as wanting to know what I could do to help, like many who have come "late" to animal advocacy, I also wanted an answer to this question: *Why was I asleep to the horrors for so long?*

It is easy to blame the media, the government, and the meat and dairy industries for their (incredibly successful) marketing of consuming animal products as a natural and necessary behavior. It is more difficult to look at the subtler ways in which we are socialized into "eating our friends," as sociologists Matthew Cole and Kate Stewart succinctly put it. As Brian Luke has said: "Enormous amounts of social energy are expended to forestall, undermine, and override our sympathies for animals" (106). It is even harder to turn to one's bodily practices. What was it *in me* that held me back from seeing what was being done to all the nonhuman others?

This was not an easy question to ask, and I did not willingly face up to it. But asking this question was a way to connect my body with the broader "energies" that govern how our social world works, including how it undermines and overrides our sympathies for animals. Considering all this, perhaps a better question might have been: *How did I finally wake up?*

Since our social world is structured by our bodies, the answer to this question was to be found in my body. After a lifetime of repressing my emotions, I was in such a poor state that I had no choice but to open up my body of work, to see what was written there.

Thirty-five is the classic period of midlife that encourages reflection on the future. As Dante writes at the opening of his *Inferno*, it is the age at which we may find ourselves lost in a dark wood, our path ahead uncertain. On the face of it, my life appeared successful. I'd worked and lived abroad, traveled widely, and committed my time to campaigning for causes to reduce injustice and suffering. However, this did not give me enough time to write for myself—literally, the need to write to *be* a self, writing being the protective armor that helped form my identity.

A simple moment of decision that precipitated the overhaul of my life arrived a few years earlier, when I was offered a more senior role at a nongovernmental organization in Palestine. I'd studied the intractability of the Middle East crisis and I had a yen to go to what I saw as the source of the world's problems, to make my imprint. The role involved management–donor relations and travel, but not much writing. I didn't take the job. I was living in Italy at the time but resigned from my role as an editorial project manager at a media charity and returned to the United Kingdom. I moved into academia first to study and then to teach journalism and creative writing as a junior professor. I published scholarly articles and began a novel. My intellectual future seemed secured. Perhaps this was no surprise, considering I'd privileged its success for thirty years.

Yet nothing *felt* successful. I knew I'd not made the most of my time or energy. I'd bounced from opportunity to opportunity through my twenties without making a full contribution, mistaking restlessness for spontaneity. I blamed all of this on a tug-of-war between the self-indulgence of wanting to write novels and wanting to be out in the world righting injustices. In academia, the frustration was the same. Deep into my thirties, I came to see that it wasn't the situations that were unrewarding; it was simply that I couldn't focus. I came to the obvious but shocking realization that the common denominator in my unfulfilling experiences was myself.

Nowhere was this clearer than in my personal life. In April 2010, aged thirty-five, I ended what was cursory, painful romantic relationship number thirty-one. *Thirty-one.* None of them, obviously, lasted very long. My emotional health—the life of the sensing body—was a wreck. I thought I would never be able to maintain intimacy. I wasn't sleeping. I was in debt. I called myself a "vegetarian," but I wasn't healthy. I still sought out comfort foods that signified parental love, including bingeing on chocolate and secretively eating fish, even meat. I was often ill or injured, pushing my body through running long distances and working long hours. My digestion was poor and my immune system was collapsing. The therapists I saw were as short-lived as my relationships. After heading to a meditation class following my final breakup, I realized I couldn't even breathe properly. My inhalations were ragged and shallow; my exhalations hurt my throat. If this most fundamental of all acts was so difficult, I thought, then what hope was there for everything else?

My body was coming apart. For most of my life I'd pulled myself together just enough to keep going. But all of that "keeping going" had not really gotten me anywhere. I'd reached a crisis point; I could go no lower. Not only did my behavior in intimate relationships need to change, but, in a sad parody of the world's larger problems named in Naomi Klein's book, *everything* needed to change.

As he explains in *The Body Keeps the Score*, psychologist Bessel van der Kolk shows it is our corporeal body that does indeed maintain a tally of what happens to us: for example, the manifestation of my childhood stress as lifelong insomnia. Trauma is trapped in the body—and I'd kept mine imprisoned for decades. At his Trauma Center in Boston, van der Kolk and his team treat war veterans and victims of incest, abuse, and rape with massage, yoga, and eye movement therapy, as well as traditional talking treatments. Mindful breathing is the basis of their work to relax the sympathetic and parasympathetic nervous systems, especially the vagus ("wandering") nerve that connects the internal organs to the central nervous system. It is with bodily therapy that the individual can "switch on" those parts of the body–mind that are "knocked out" by serious trauma or "closed down" during persistent low-level trauma. This "switching on" helps the individual process traumatic life events into a narrative, and such a narrative remobilizes the body into understanding it can escape from its situation—just like our pig in thin air, whom we are soon to meet. Without this processing, the trauma gets "locked in," as activist and professor of psychology and sociology at the University of Massachusetts Melanie Joy describes in her experience of killing a fish at the age of four, in the collection *Turning Points in Compassion*: "Traumatic memories are unique. They are immune to the passing of time and remain eternally present, evoking images and feelings as clear and as powerful as when the event took place" (141).

For most of my life I didn't want to know how I felt—or, as a verb, *how* to feel. I'd created a false self instead. Psychologist D. W. Winnicott introduced the phrase "the false self" to describe the defensive façade of an individual forming an identity that would keep the ego safe from dissolution. Such a tactic leaves the person lacking spontaneity and feeling empty because of the rigid nature and overbearing energetic demands to keep up this fabricated picture. These false pictures require the repetition of behaviors and cognitive reinforcement to be maintained. Change the behavior, and the identity will also change. However, we become attached to these false selves because we believe they "are us," even when we understand how destructive they are.

We've come to understand the idea of the false self as a mask. For Chris Shilling, the masks we put on—generally successful, high-achieving, individualist images of identity, but also masks of invulnerability to loss, grief, and care—are indicative of "who controls the norms against which individuals mould their bodies" (226). Masculine identities, for example, are worn as masks and illustrate the power of these high-achieving, individualist, tough, and unfeeling ideals. As is obvious, not all masks are worn through coercion. Although they take their own toll, no one appears to benefit more from the masculine masks than the men who wear them. The mask often looks like it meets one's desires. What was *wrong* with being a white, educated, male editor for a global media charity, or a professional, published scholar?

As Shilling says, what's important is the location of where identity masks are worn: the body. My body was telling me my way of experiencing the world was wrong—the symptoms were there for all to see. If the self we project outward from the body is inauthentic then the body will suffer, regardless of whether or not the chosen mask appears successful. This is why it took me so long to accept and allow my compassionate entanglements with other bodies. For the majority of my life, my body was the last thing I wanted to think about. I was not "thinking through the body" (3), as Lisa Blackman suggests we must if we are to reimagine the frameworks that organize our experiences, such as the relationships between the individual/social, structure/agency, mind/body. We need to "turn on" again the processes that have been turned off, including our sympathies for nonhuman animals. Only then is what we might become a "becoming-with the nonhuman" (277), says Kate Wright, a "becoming-with" not only animals but the very air itself—for example, when our hair stands on end just before we're about to get struck by lightning. We need to relearn what we have learned to suppress. And this takes effort.

Learning How to Sleep

I had to change; there was no alternative. The first bad habit to ditch was destructive relationships, and I stepped back from romantic love. I persisted with mindfulness classes to learn how to breathe. Perhaps the most pressing issue was my lack of energy. If my body was not rested, then I had no hope of ever making major changes. So my first painful task was to learn how to sleep. Perhaps more than any challenge over the past five years, this one gave me insight into the knowledge that if we want to have new and compassionate relations with nonhuman animals, it often begins with changing not what we think about *them* but who *we are* in our bodies.

By 2010, I'd not slept a full night's sleep since . . . well, ever. I didn't know what it was to experience a night where you fell asleep at 10 P.M. and woke at 6 A.M. In fact, the idea terrified me. What would happen as I slept? Hypervigilance was a habit gathered from childhood to keep me safe, but it was now redundant and damaging. I was exhausted, living far below my potential. So I enrolled in a sleep research study at the Glasgow Sleep Center. I hoped I would change my poor habits. What I didn't know was that I was about to change my identity.

Aristotle tells us "we are what we repeatedly do." He was talking about the fields of excellence, but the idea also holds for our negative habits. If what we repeatedly do is not finish things or break off relationships, then we are what we repeatedly do *there*, too. Our identities are born out of repetition. As philosopher Judith Butler has taught us, repetition is a form of power: have people repeat something enough (such as eat animals) and they become identified with that behavior. Even if they come to see that action as damaging, it is a part of who they consider themselves to be in their social world, which approves of such behavior. Giving that up is painful.

Sleep scientists know already what makes positive change in our sleep habits: the ability to stick to an acutely painful program will be the marker of successful change. When people are suffering from insomnia, the program has you get *less* sleep. It forces you to improve your sleep by first improving the quality of that sleep. And on it goes, slowly, painfully, over a three- or six-month or even three-year period.

The body takes a long time to reset its biological patterns. We had to realign the homeostatic and circadian sleep systems so they could work again in sync. It was physically, outrageously painful. In my first week, I hallucinated. In the second, I stabbed myself in the hand de-stoning an avocado and needed five stitches.

But I stuck to the program. Around eighty percent of gains in sleeping are made through behavioral—

embodied—changes, such as keeping the same rising time, abstaining from coffee and alcohol, and not staring at screens before bedtime. However, a cognitive element exists, too. You have to change your opinion of yourself as a poor sleeper. You alter your *sleep identity*. Without that change of identity, even if the sleep improves the old habits are likely to slip back in.

As I altered my sleep behaviors, I was able to challenge the idea that I had of myself of sleeping in a particular way (poorly). One might imagine that sleeping is a natural behavior; I discovered that it was learned. Our learning begins as infants, and not in our own bodies but through regulation from the bodies—the heartbeats—of caregivers. Because my identity as a poor sleeper was learned it could be unlearned. My sleep improved. I woke less often in the night and my sleep became more automatic. Indeed, I no longer thought about sleep all the time, and especially how little I had.

I could also see what false protections a sense of myself as a poor sleeper had given me. It provided me with excuses not to commit. I could never dedicate my time to writing after a poor night's sleep. But was it the other way round? Was I so afraid of failing at something I longed for that I created an excuse for myself? Did my identification with insomnia let me off the hook in relationships, too? The answer, sadly, was *Yes.* I created and then used my self-identity as a poor sleeper to protect other parts of me that I was not able or willing to show people. With poor sleep, there was always an excuse not to be my (best) self.

Learning about how I pushed away the world by inhabiting an always-tired body led me to understand that if I were to care better for my body, I would change my relation to the world. This improved self-care meant a process of relearning to allow one's affective responses to emerge as emotions—to pay "attention to the particular" of my own body and to act on those messages. Badly schooled in relationships with people, I drew upon a small handful of successful templates—which included the response, half a lifetime before, to the slaughtered whale. I sought out voluntary work with a marine conservation charity, ORCA, educating schoolchildren on cetacean protection. I helped young people write poetry, stories, and reports about our entanglements with marine animals—in terms of fishing lines, sometimes literally—and came to care for the bodies of whales once again. My behavior became that of an educator for animals. Slowly, my identity shifted, too.

Giving what Iris Murdoch has called "loving attention" to my own body was made possible by learning how to sleep. New identities as an animal educator *and* a good sleeper revealed to me an ability to feel and act upon my entanglements with others, including within the animal agriculture industries. To put it more bluntly, to wake up to these horrors I first had to learn how to sleep.

Henry David Thoreau wrote in *Walden*: "Moral reform is the effort to throw off sleep. . . . We must learn to reawaken and keep ourselves awake, not by mechanical aids, but by an infinite expectation of the dawn, which does not forsake us in our soundest sleep" (65). Learning to sleep was a bodily transformation that allowed the moral awakening Thoreau speaks of, and this brought my moral attention to the bodies of others. I found out that it was possible to recalibrate how I existed in the world by moving out of my head and back into my body. I stopped trying to intellectualize my feelings and actually *felt* them. I discovered how to read my body again, after thirty years of not doing so. It was painful, but worth it. As Carol J. Adams told *Satya* magazine, "even consciousness about suffering is a gift. We experience the suffering but we are also given the gift of the consciousness about it, and it's better to be awake than asleep."

From Sleeper to Animal Advocate

As the animal advocate and vegan author Colleen Patrick-Goudreau puts it, we do not find our compassion for animal others—we *re*-find it, after it has been socialized out of us, or "turned off" in van der Kolk's terms. Patrick-Goudreau's assertion is that as children we are inherently compassionate for the world around us. Brian Luke agrees, arguing that sympathy for animals is a deep, primal disposition repressed out of us as children by energetic social conditioning—conditioning that is, in its own way, traumatic. Our entanglements with nonhuman others are strictly controlled and enforced during childhood, and channeled down culturally defined routes so that we begin to see, again through repetition, our relations to nonhuman others in very narrow terms: as food, as pets, as vermin.

Our social identities are thus powerfully defined by our caregivers and social world. As Shilling has already referred us to, our identities often fit with society and culture in ways that make those fits comfortable and pleasurable, and we value our identities shaped in this way—we are recognized for them and they are rewarded. Indeed, we might need some of these identities to buttress a sense of self that would fall apart without them. Lori Gruen notes that, especially for those whose "subjectivity, agency, and experiences have been undermined, questioned, or denied," such as those who have suffered trauma or stress, difficult childhoods, prejudice, or discrimination, "the maintenance of a self-identity is an achievement and not one that they are willing to give up so readily" (62). The irony is that considering that many of our Western social structures are traumatizing and discriminatory, this could include most of us.

What is crucial here is the idea that, as Gruen says, "[t]he distinction between self and other isn't one of distance and it doesn't entail dominance or subordination; it can be maintained in ethical ways. It is centrally important that one has a balanced and clear self-concept to be able to engage empathetically with others" (62). What Gruen argues for is essentially the experience I underwent. I was able to regain a "balanced and clear self-concept" by getting back in touch with myself as an embodied being. As such, I understood how it is that identity is shaped through embodiment, not through Cartesian separation of mind and body but through regaining a "balanced and clear self-concept" to properly feel entanglements with others.

So, perhaps it is no coincidence that I (re)discovered animal advocacy as a life's purpose at the same time that I recovered my bodily senses. It is not a direct causal link, of course—if it were, our work as animal advocates would be much easier! Many people who re-find their bodily balance after trauma may not become vegetarian, let alone vegan or an advocate for animals. I obviously had some natural sympathy for animals through my childhood, and that found an initial outlet in vegetarianism. But in adulthood my stunted embodied practices left me corporeally and constitutionally too blocked to go further in feeling for others. When I was able to reconnect with my body, it gave me the capacity to live a more compassionate, feeling life.

We are always already in brain–body–world entanglements, as Blackman says, and we can develop our ethical skill set to see these entanglements through moral attention, as Gruen asks us to. For myself, I had no chance of discovering these experiences without first becoming reconnected with my body—to become "balanced" with a "clear self-concept."

My corporeal change came not through a psychoanalytic "coming to know" of a repressed past, but rather through what Elspeth Probyn has called instead a "coming to care" (291). This is an important concept for animal advocacy, a return to our compassionate nature, a care that, says Melanie Joy, "the dominant, meat-eating culture breeds out of us as we become 'mature' adults" (142).

When we "come to care" it changes our capacities for action. And a person's capacities for action are clearly relevant for anyone working in animal advocacy. As Shilling says:

> Social change does not happen automatically, and nor does it occur simply as a result of purely intellectually motivated actions. Instead, people's experiences of, and responses to, social structures are shaped significantly by their sensory and sensual selves. These variables are important as they can exert an important impact on whether people feel at ease with, and tend to reproduce the 'structures', 'rules', 'resources' or 'social fields' they are most familiar with, or emotionally experience these structures as unpleasant, undesirable and worthy of transformation. (256)

"Coming to care" implies movement, an arrival or advance. At its basis, many of the advances toward social and animal justice over the past two hundred years—through abolitionism, the civil rights movement, feminism, the environmental movement, and animal advocacy—have been made through fighting against norms and institutions that ill-govern our bodies. These movements have paid at least some attention to this understanding of how each of us is, as Gruen says, "a particular embodied being who organizes her perceptions and attitudes, a self" (66).

Does animal advocacy have the most difficult task? We are fighting not only for our own bodies, the experiences of which we are able to authentically communicate to others. We are also battling to communicate the bodily experiences of other species and why these should matter for changing our currently exploitative norms. Elspeth Probyn asks: "What forms of care are most effective for changing our behavior? How do we come to care? And which types of care have the desired effects?" (296). Lack of attention to bodily entanglements has given us a social world built upon the exploitation of other species. We need different types of care, based on different feeling foundations.

These different kinds of care are emerging. And they are precipitating a global identity crisis in all forms of exploitation. Such exploitation is becoming much more visible. It is visibly becoming a threat to our individual and species existence, through bodily illnesses such as heart disease and some cancers, and through the impact of intensive animal agriculture on our communities, environments, and planetary home. Climate change is one real, observable outcome of our separated and dualistic identities as exploiters of Earth resources, nonhuman animals, and of each other. Old repetitions are breaking down and being challenged at the level of the body. Yet even as we come to understand how our identities are formed and how destructive they can be, as merely repetitions of poor and unnecessary behaviors, we find it painful to change.

So we've come closer to my body. But what about the pig of the title?

It's time to meet her.

2

Social Media, and Becoming like a Pig

THE PIG HANGS motionless, caught in midair at ninety degrees to the ground. She looks as if she's running down the side of the truck from which she's just jumped, although she's leapt a little way out, like a diver worried about rocks. The sky behind her is a cool blue with summery clouds that float over a wide, urban street with trees lining the pavement. If it is rush hour, then it is a relaxed one in the city of Foshan in central Guangdong province, China's most populous region, with over one hundred million human inhabitants.

Alongside the truck is a silver minivan. Both vehicles are waiting at a junction; an earlier photo, where the pig is clambering over the truck's side, shows more traffic. It is not clear in that picture that the crisscross pattern over the top is barbed wire. In the second picture, when the pig is outside and falling, the barbed wire is more prominent, as is the face of a second pig looking back at the anonymous photographer. Now our escapee is heading toward the tarmac, snout first but head up, ears back, legs in a strangely relaxed, dressage pose, as if she is savoring her freedom. If joy comes not in the great achievement of plans but in the momentary awareness of one's existence, then our pig is experiencing, most likely for the first time, this wonder of the present without a thought for the future. Not how hard the ground will be underneath her; nor what might happen if the driver realizes he has lost one of his "units." In this instant—caught in the air of a summer's day—she has a glimpse of a different life; of autonomy over her body; of freedom, even if it only lasts a moment.

—

I first saw this picture in June 2014 via Facebook, circulated by one of the advocacy organizations that share such stories. The image arrived at a time when pigs had become central to my thinking on the connections between species. There were links between my sense of justice, life purpose, male role models, and the pig that I was just beginning to unpick. It arrived at a time when I was also thinking about images: their role in my journey toward veganism; how bearing witness to the reality of life for farmed animals was part of the process for changing my relationship to all species. For some, however, images are as close as we get.

When I first saw the picture of this pig, I followed the progression of quick snaps taken by a passenger in the following car: from the moment of the pig's outlandish leap to that of her safe standing. And then, for the time being, the story ended. The original reports suggested, as in most cases of an escape, that the driver rounded her up and put her back on the truck. Only the very few are fortunate enough to survive the leap and escape, and then negotiate a route away from traffic, trigger-happy police, and scavengers. I feared for what would happen to this pig who'd seized her moment to free herself from her confinement in the back of a truck on the way to slaughter. And then my Facebook feed refreshed.

Haunted Humans

Mark Hawthorne, author of *Striking at the Roots* and *Bleating Hearts: The Hidden World of Animal Suffering* (whose front cover depicts a pig looking out of a transport truck, a photograph by Jo-Anne McArthur), has an image of a single animal body that haunts him, too. His was in an advocacy campaign email. It was, he recounts, "a gruesome image that showed the bloody bodies of dozens of freshly skinned seals scattered across the frozen landscape. Near the bottom of the photo, pondering this horrible scene, was a lone seal who had managed to escape the carnage. . . . This was the photo that occupied my consciousness and kept me awake at night."

John Sanbonmatsu, Associate Professor of Philosophy at Worcester Polytechnic Institute, Massachusetts, introduces the edited collection *Critical Theory and Animal Liberation* with another. The image is of "a group of Midwesterners standing in a circle in the snow, cheering on a young boy of about seven years old as he beat a fox to death with a baseball bat. The boy, with a bright smile, stands with his legs firmly planted, as though waiting for a pitch that never comes. The fox, crouched, tongue lolling, exhausted almost to the point of death, gazes vacantly, a look of hopelessness or resignation visible in his pinched face" (1).

Perhaps we all have that one image that keeps us awake. Some will have multiple. Some have chosen to look away. In my case, the image of a pig hanging in the air became another in the photo stream of consciousness that I drew upon on the journey toward animal advocacy. The story circulated around the world, as tales of escaped farmed animals do, because we like to feel compassionate about individual animals when they show a desire for life and for a moment become, as we imagine ourselves to be, exceptional beings with purpose, who understand the meaning of freedom. Many books about animal advocacy that tackle the subject critically, as the two examples above show, begin with an image. An image makes the individual for a moment visible—a counter to the incomprehensible tally of farmed animals we slaughter in a process that Barbara Noske calls the "de-animalization" of the animal self: from sentient being to packaged product. The individual nonhuman animal disappears behind the sheer amount of bodies. Animals may, as Claude Lévi-Strauss believed, be good to think with, but "numbers help us to stop thinking," say the scholars Vinciane Despret and Jocelyne Porcher (36). When an individual breaks out of the mass, it helps us think again. As British geographer Henry Buller puts it, "you do to a pig in an intensive farm what you would never do to a pig if you had only one or two of them" (162).

The image of the pig suspended resonated with me (and many others, apparently) for a number of reasons. When we see the intact bodies of pigs, if we see them at all, they are often hanging in this way at ninety degrees, but usually dead or dying, and if dying then painfully so, waiting to be dipped in a tank of scalding water to remove the bristles from their skin, held up by one back leg on a moving chain. These are the depictions in artist Sue Coe's 1989 painting *Union*, where pigs form the backdrop to the removal of unionized men from their low-paying jobs in the Chicago stockyards—both human and nonhuman bodies fungible parts in the processes of production (a situation most famously described by Upton Sinclair in his 1906 novel *The Jungle*). The suspended animal body has also become part of our cultural lexicon since artist Damien Hirst put a fourteen-foot tiger shark and then cows and pigs sliced in half into formaldehyde (for example, *This Little Piggy Went to Market*, 1996) and exhibited them for our entertainment and, possibly, our education.

So seeing this pig alive, even if in a picture, resonates with but also troubles those images we know of pigs hanging. Because this is an image of life, not death. The photographer has captured what Virginia Woolf referred to as "moments of being"; one of those "intimations of immortality" (Wordsworth this time) that make us recognize we are alive, and that this itself has meaning. Perhaps it might make us ask why pigs need to be dead for us to see them this way, in their "moment of being." That is, do pigs only have meaning (for us) when we end their lives—is this the only important "moment," when upended? The pig is one of the few farmed animals whose body parts or secretions we don't use before her death, for her milk or eggs or wool. Are we indifferent to her momentary, joyous existence? The pleasure of farmed animals is rarely, if ever, considered. *Our* pleasure, however, seems to be our central preoccupation. In Sanbonmatsu's tale of the boy bludgeoning the fox to death, the adults accompanying the child are standing around "grinning. And it is this last detail," says Sanbonmatsu, "of ordinary human beings taking delight in the torture of a powerless individual, an animal, that still troubles me the most" (1). This pleasure in cruelty scares us. As philosopher Cora Diamond says of J. M.

Coetzee's character Elizabeth Costello in *The Lives of Animals*, she, and we, are "haunted by the horror of what we do to animals" but we are equally haunted "by the knowledge of how unhaunted others are" (46).

To be haunted is to be unable to shake off an unsettling feeling, to be touched by someone whom we believe is no longer there. To be haunted is troubling because it is an embodied experience but one that carries with it the unease of *dis*embodiment. The ghost—whom we cannot shrug off—is a memory of a previously alive being whose spirit lingers, as she did not die peacefully. To feel haunted is to be reminded of the evanescence of bodies and to be made uneasy about the sense that the embodied being who is now dead was once entangled with us, and so cannot, while we live, ever fully die. The ghost is unquiet, and those of us who are haunted hear its claim to life, its wish to return to its body from which it was violently put out.

Perhaps what troubles us in such images is this inability to stop what has already happened. The image is always past; in most cases, the cruelty has already been inflicted; the animal is already dead. Although the work of activism focuses on what we might halt in the present and put an end to in the future, it is the past, the revenants and ghosts captured in these images, that haunts us. They threaten our actions with a sense of meaninglessness. As the geographer Tim Ingold puts it, we are "haunted by the spectre of the loss of meaning that occurs when action fails" (80). For Hawthorne and Sanbonmatsu, the images are both open wounds and open questions about our capacities to act. Through the power of social media networks, animal advocacy organizations are regularly circulating images of nonhuman animals to bring us closer to the reality of their lives and the ugly injustices of our own. But asking us to act is also asking us to be haunted, to take upon ourselves what philosopher Lisa Tessman has called the "burden" of compassion. As I stared more deeply into this image of the pig escaping the transport truck and moved to engage further with animal rights online, there were ever more opportunities to be haunted.

Learning to Love Social Media

Our contemporary communications practices are dominated by visually oriented social media. Facebook, Twitter, Pinterest, Instagram, Behance, Google+, Snapchat, Blogger, and WordPress are some of the ways we gather our news, stay connected to friends, contribute to groups, organize our visual catalogues, and express our identities to ourselves and to others. The mobile is globally ubiquitous, from Barack Obama and his BlackBerry to the homeless woman on the sidewalk outside my office using her cell phone to coordinate a night's sleep once she's raised money for safe shelter. More and more online content, especially social media content, is being accessed on our phones. We now do not "look" at our phones, but rather experience what the advertising agency Universal McCann has dubbed "mobile moments." It is almost certain the images of the pig escaping the slaughterhouse-bound truck were taken, and immediately uploaded, via a mobile—a word that itself, ironically and appropriately, summarizes her sudden release from confinement and lets her hang forever.

The brightest organizations, networks, and collectives understand social media and use it to achieve their goals, whether to topple dictators or win elections or simply build a brand. Animal rights activists were among the first to embrace social media. People for the Ethical Treatment of Animals (PETA) was placing videos on its website as early as 1997 and building forums for its young members through peta2.com not long after.[1] Animal activism has embraced social media specifically. Everything is now social. As *National Geographic* noted in 2014, animal rights organizations such as PETA are claiming that social media is having a massive impact on how people view nonhumans.[2] In 2013, over forty-five percent of YouTube users had posted videos of pets or animals on the site. In 2014, one in five YouTube users globally between the ages of sixteen and thirty-four were regularly viewing videos of pets and animals. In general, people are watching for longer, too: in July 2015, YouTube watching was sixty percent longer than just one year before, driven mostly by people having "mobile moments." Some studies even suggest that looking at pictures of animals at work can make you more productive.[3] Of course, not all of these are animal rights videos, or show the inside of slaughterhouses. Yet in response, for example, organizations such as the Global Dairy Platform and International Milk Promotion Group are looking at "the challenge of the anti-dairy movement in social media" as a credible threat to their business.[4]

In the United Kingdom, the largest animal rights organization, Animal Aid, won the 2014 *Web User* magazine award for "best new website" for its campaigning tool identifying how politicians voted on a number of animal-related bills. The website is linked to Twitter, Facebook, and email to help spread the campaign, and Animal Aid has over 110,000 followers on Facebook and 55,000 on Twitter. These U.K. figures are dwarfed by those for American-based organizations. The networks of followers for Mercy For Animals (MFA) in the United States stands at the time of writing at 1.4 million Facebook friends and 140,000 Twitter followers. MFA focuses as much on media as on message and is sophisticated in its research-led analysis of how the two interact.

My first exposures to the vulnerability of living animal bodies came almost exclusively through social media, via Facebook and Twitter. Beginning around 2010, I was able to seek out the organizations working for animal activism and liberation, and the materials they draw upon: many horrific and graphic, some pleasurably joyful, all emotionally affecting. I cannot recollect my first encounter, but I do remember my own "moments." I recall watching the six-minute meat-animal clip from the film *Samsara*, which went viral just before the movie's release in 2011.[5] I remember discovering Toronto Pig Save, seeing some of their non-graphic imagery—photos and video—and being moved to tears. I joined groups online and began to connect with others facing the exploitation of animals.

I moved to become vegan because of this imagery. I soon reached a stage of my transition where, feeling strong enough because of the bodily changes I'd made, I wanted to push myself into a state of new vulnerability. I didn't know at the time that this was not about vegan food practices as such, but about going beyond into advocacy—a way for my (re)emerging feelings of species injustice to find an outlet through action. I realized I wasn't connected enough to the reality of the lives of animals who were suffering. I was distanced from the other; my consciousness of their vulnerability was too far from the particular and individual. It was not that there were no ways to contribute—I just couldn't find my way toward them. There was something I had to break through.

So one evening after a night out with friends I forced myself to sit down and watch footage of pigs being slaughtered. I wept loudly and painfully. I cried for my friends whom I'd been to dinner with and who'd eaten meat and didn't want to talk about the issue. I cried for myself: it was my moment to fully awaken to the horror of what I'd been ignorant of in my life so far. I cried for the individual pigs. I knew I'd have to see their suffering directly. I also knew it would be easier from then on.

That evening was a turning point made possible through the networks that had educated me to fully face the issue. The tissue of "social" and "media" came together to provide a psychological environment of readiness to face the imagery, if not a physical community. Through social media, I felt connected to others feeling the same.

—

This was my initiation as a witness. To bear witness is to apprehend, whether one wishes to or not (consider the bystander witness), the truth of what is happening during an event. It is to open one's eyes to what is taking place, but not only one's eyes. Without an individual caring first, all the video footage in the world will not change his or her worldview. Melanie Joy suggests that a turning away from care is the psychological act performed by meat eaters: it is because they *do* care that "they feel compelled to turn away from the atrocity" (145) of the exploitation of animals by ignoring or rejecting calls to stop eating meat. My initiation into the world of nonhuman animal suffering was accelerated through bearing witness (online) to scenes of animals being put through the processes of exploitation and slaughter. I was also bearing witness to others bearing witness: videos of the vigils outside slaughterhouses of the Save Movement and of the silent vigils organized by Animal Equality, where groups of activists occupy city center locations and hold the bodies of nonhuman animals to recognize and honor their deaths. I will have cause to talk more about these organizations in later chapters.

Witnessing is a strange act, what Shoshana Felman and Dori Laub have called a "nonhabitual, estranged conceptual prism" (xv) through which we apprehend and make available to the imagination of others the events that are seen. Felman and Laub's work arises out of the witnessing and testimony of Holocaust survivors, where the task of witnessing is to account for "the scale of what has happened in contemporary history" (xv). One could argue that the scale of what has happened in the history of nonhuman animals requires us to cultivate a similar "nonhabitual, estranged conceptual prism" to account for all that has happened. But that is not the aspect of witnessing that I want to focus on here.

What is the value of bearing witness? In her book *Political Emotions*, the cultural and political theorist Martha Nussbaum puts forward the idea that "good societies" should cultivate in their citizens particular emotions, such as compassion. This cultivation of certain emotions provides for stable and motivated societies. For Nussbaum, this is not to advocate for emotions because they would somehow be infallible markers of correct public policy. Emotions, rather, have both cognitive and evaluative elements that are, or should be, in dialogue with the principles of society.

One of the main ways to foster compassion in society, argues Nussbaum, is through the concept of "tragic spectatorship." She states this was a beneficial aspect of ancient society, especially in Greece, where the spectatorship of "tragic" Greek theatre "dramatized the psychological and bodily suffering and tragic choices" faced by its citizens, and, says political theorist Alison McQueen, "thereby [allows] the spectator to imaginatively enter into the world of others" (654) and consider those difficult choices for himself (the audience would have been almost exclusively male). Such tragic spectatorship, argues Nussbaum, contains powerful narrative content that educates the body emotionally as well as the mind politically.

McQueen is critical of Nussbaum's argument. She suggests that Nussbaum does not make the case strongly enough—that Nussbaum "avoids the more difficult task of demonstrating that 'tragedy undermines exclusion'" and, as McQueen goes on to say, "this argument about the political effects of tragedy is, at best, suggestive" (654).

Yet, I would argue, both Nussbaum and McQueen miss the point of what is valuable and effective in the idea of "tragic spectatorship." Both concentrate on the content of the tragic plays rather than on the active aspect of spectating—which is a phenomenological process of witnessing. That is, it is not only tragedy that brings us close to compassion for the other; it is the *act* of witnessing that tragedy in person and with others. If a tragedy educates its witness to the "psychological and bodily suffering" undergone by those actors in the tragedy, then its value is in the witness empathizing—*feeling*—those psychological and bodily traumas suffered. The witnessing is a performance in and of the body. The "tragic" refers both to the image seen and the seeing—and this becomes a useful concept for thinking about the process of bearing witness in animal activism. In this sense, bearing witness to the suffering of the animals whom we see for just a moment in the transport trucks on their way to slaughter, and of the videos of those last journeys, are acts of "tragic spectatorship." But do they educate us in forms of political emotions, such as compassion? And does this education allow "the spectator to imaginatively enter into the world of others" and act accordingly?

My speculative response is *Yes, they do and it does.* I did not look away. But is this the case for everyone? What other factors need to be in place for the videos and images circulated via social media to have such an effect? And why did this image of the pig escaping have such an impact upon me? Perhaps it was because it was one of the images I could stand to look at for so long, and for this reason it was able to lead me to a "coming to care." As Susan Sontag warns us, the omnipresence of disturbing images desensitizes our capacity for compassion. Sontag argues that "a pseudo-familiarity with the horrible reinforces alienation, making one less able to react in real life" (41). This had been my experience of street protests after all. I felt that the animal rights protestors I met at the demonstrations were alienating themselves from the people they were trying to reach, turning people away from witnessing with images that were too graphic for them to bear.

Jodi Dean, a political social scientist studying the effects of social media, believes such alienation online has much to do with the structure of the ways in which we receive, share, and monitor imagery. For Dean, each Tweet or Facebook post is a tiny "affective nugget" that catches people in a loop of reproduction that distracts them from real political work. It is addictive because it is affective, and stimulates in us a charge of desire—of wanting to see and be seen. In other words, the energy does not come so much from the imagery or messages we share, but through *us sharing them*. Each affective charge operates through a process of repetition to ingrain in us a desire for the next hit, which tells the reward centers of our brain that this behavior is good. It's a bodily experience, and the actual content of the imagery becomes less important to us.

If Dean's analysis is correct, then the problem with this scenario is political and economic. While this structure of how social media operates may be individually *affective*, it is also politically *ineffective* for systemic change. Rather, social media platforms remain traditional in the political economy of contemporary life in that they make huge amounts of money for a small number of owners and shareholders. They do this through selling our leisure time and social interactions, including our activism, as consumable spaces full of advertising, where we are tracked and profiled often without our knowing. And while we are spending our time being social on media, argues Dean, in becoming affectively charged, we are not challenging the status quo that continues to divide the world unequally.

Social media is not used only by animal advocates. In 2015, Ontario dairy farmer Andrew Campbell (@FreshAirFarmer on Twitter) launched the hashtag #farm365 to show the "true face" of animal agriculture.[6] According to *Real Agriculture*, Campbell's project aimed to "shed more light on the daily happenings of a farm" and "at least in part, to spur on conversations around food production."[7]

For animal rights activists, the project was propaganda for an exploitative system that shows a sanitized version of farming, and the hashtag was flooded by advocates with counterimages of violence and exploitation. A few days after launching the #farm365 project, Campbell wrote a piece for the *Let's Talk Farm Animals* blog. He accused animal activists of posting doctored photographs on Twitter as a way of cheating the public: "Their truth usually involves Photoshop, graphic images and misconceptions of what an animal needs and wants."[8] For Campbell, his turn to Twitter has been a form of "activism" calling people to his aid. Being under attack, Campbell raises his hat to his fellow farmers, who all heard his call: "And when one of their own was on the brink [he doesn't say of what], they came running. Many showed their farms in the moment, sharing their beliefs and systems. . . . A look at what really goes on behind a barn door or in a field. No Photoshop, no terror, just fact."

Campbell's insinuations that animal activist images are doctored (not "fact") and his over-easy fallback on "terror" are stock responses to what the farming industry feels are unprovoked and unjustified attacks. The law, too, is generally on their side, especially given the Animal Enterprise Terrorism Act (AETA) and ag-gag bills. The criticism that Campbell makes of the images circulated on social media by animal activists is the same criticism that can be leveled back at Campbell: that of selectivity. The images from farmers on the #farm365 stream show none of the intensity, high mortality rates, stress, enforced impregnations, removal of young, and slaughter—all of the things we can guess our pig was escaping from—of intensive agricultural practices, which still account for ninety-nine percent of all animals farmed in the United States (and much of Canada). The imagery also lacks the impact of the other senses: the sounds and the smells of the conditions in which animals are kept. As Howard Buller asks: "How do we become affected by these farmed animals? How do these animal lives—lives that we violently end for our own purposes—matter?" (160). There are limitations to the image, and audio-video to a lesser extent, in how they affect us.

#farm365 also shows, with Campbell "on the brink" and defended by the metaphorical bodily actions ("they came running") of others on his "side" against the "terror" of the animal activists, how strongly people will defend their identities when they *feel* corporeally threatened, even if there is no real physical threat. In a way, however, the feeling is very real. Identity formation and defense are bodily processes. An attempt to change someone's beliefs is also an attempt to change his or her behaviors. Beliefs are *felt*—and often felt to be pleasurable. To attempt to convince someone to go vegan means a change in that person's bodily repetitions. As most of us know, that can be a painful process, especially when those bodily acts are social responses (saying *no* to family food rituals, for example). The activist Gay Bradshaw writes of her own conversion: "Unconsciously, I felt that becoming plant-based comprised a break from my familial customs, which meant a loss of connection with those I loved. The cultural rituals associated with my family life were integral to self-identity. Giving up these rituals seemed like a betrayal" (55). I come back to these points in more depth in the next chapter.

As the Arab Spring and other events have shown us, social media is increasingly *not* the channel through which traditional elites maintain their ownership and control of dominant messages. Nor is social media much used in the way Campbell wants—as unchallenged forums. When large corporations step into social media they are often clumsy and out of step with popular commentary. There is something likeably transparent about the nature of social media, whether that be a prioritization of the wisdom of the crowd or the low barriers to entry. Voices are not silenced by gatekeepers—at least in most countries.

Yet for most people living Western, urban lifestyles, such an image of a pig jumping from a transport truck, caught on camera and circulated via social media, will be the closest we come to seeing the alive body of a pig. And that is how many would like to keep it, not out of ignorance but out of fear. If we never come to know the lives of the pigs and what they experience, how can they move us?

As it turned out, this image of a pig leaping for her freedom *did* move me. I continued to bear witness via social media networks: to face the pig as if she has, *pace* Emmanuel Levinas, a face. If I was to be an advocate for animals, then it was to be the fate of the pig that I would investigate; that it would be the pig whom I chose to look in the eye, promise to come closer to; that I would not turn away.

The Outsized Effect of One Animal

As I indicated earlier, it was not as if the pig hadn't made herself known to me before. Yet, for thirty years I turned away.

As a child I turned my nose up at pork chops, but my familial security and masculine identity were tied up with a desire for pig products, which were central to my formative and affectionate memories of growing up, and especially with my male role model, my grandfather.

My grandfather was from Norfolk, in the eastern part of the United Kingdom, and grew up working on farms. Before World War II he trained as a wheelwright, and then became an army paramedic. Although he moved to Kent in the southeast, he supported Norwich City Football Club and always had the Norfolk-famous Colman's Mustard Powder on hand, added to most dishes. His admonition to *Just get on with it*, his ability to delay gratification for the benefit of others, and his compassion for his cats have been models and inspirations for me to follow—although, as my history attests, I wasn't always able to *Just get on with it* (apart from befriending every cat on the street, that is). He also had other traits that I unconsciously absorbed, such as his relations to nonhumans used for meat.

Some of my fondest memories of my grandfather include foodstuffs derived from the bodies of animals. My grandfather and I had a particular ritual when my sister and I stayed over at my grandparents' house in Kent, England's "garden." He and I would wake early and walk to the butcher's shop at the end of the road to purchase Cumberland sausages and serve them up with toast and bramble jelly. The butcher's shop floor was covered in a fine litter of sawdust, and I recall the perfume of shaved wood and its feel against the red-tiled floor; the tilted glass counter filled with cuts of meat; the stripes of the butcher's apron. My grandfather would order the sausages, eggs, and a joint for Sunday dinner, and perhaps some fresh calves' liver for his cats, Scruffy and Tigger. Then we'd walk home. I cannot remember much conversation between my grandfather and me. All I know is that I felt safe in his company and that what we had done seemed right, appropriate, even essential in some way to my identification as a member of the family, first as a boy, and then a man. Later in the day, my grandfather would teach me woodwork in his shed, a place I still dream about now, even though I've not been back to the house since he died in 2005.

Although he had a great love for and patient manner with his cats, my grandfather was a confirmed carnivore. It never crossed his mind to consider farmed animals as destined for anywhere else but our plates. Considering the geographical and temporal conditions of his life, this is no surprise. But the bond I had with my grandfather was made over the bodies of those animals. Why should I have questioned the eating of animals when the person I most admired in life provided a meat-eating model?

So while pig products were visible on the plate, the life and body of the pig remained invisible. Pigs are perhaps the most hidden of farmed animals. Although a train journey between most major cities, in Britain at least, will take you through fields of cows and sheep, you rarely see pigs. They are almost never any longer, as chickens are, kept as pets or companions, although in the nineteenth century the "cottage pig" (one or two pigs kept in a sty next to the cottage) was a common mainstay for laborers across rural England. Permaculture workshops do not train people on keeping "backyard pigs." Whereas we often encounter wild geese or ducks, the same is not true of pigs, although there are wild boar in the United States, Britain, and across Europe and Australia (which many, including members of the British royal family, still enjoy hunting). Other than on city farms, where children get to see animals on school visits, farmed pigs are perhaps more invisible to us than any other land animal we eat.

Invisible yet entangled, the pig was there from the start. She accompanied me on my journey from butcher to grandfather's kitchen table, from meat eater to vegetarian to vegan. It's possible that my discovery of how pigs are treated was so overwhelming because of this emotional attachment to their flesh in the shape of a Cumberland sausage, an attachment made with and through my grandfather. To question the inherent speciesism in those memories is to set out alone on a path no longer modeled for me by this man. It has taken bodily effort to separate my feelings for my grandfather from the acts of eating meat, and to understand which elements of my inherited identity I can retain and which I can jettison. That bodily effort involved indirect encounter first through social media and visual imagery, and then directly with the actual bodies of nonhuman animals who were alive but with only moments left to live.

We are messily entangled and in co-constitution with others, human and nonhuman, grandfather and pig. There is companionship here. As scholar Anat Pick says, this "fellowship is ridiculous, ungainly, carnivalesque even—but solid and unquestioning. It is rooted in bodies exposed to time and at the mercy of gravity" (188). Even so, sometimes the longer I stared at the body of a pig, the more it vanished into thin air.

When I was at college I used to play music at parties, bars, and clubs. One time, a friend invited me to DJ at his twenty-first birthday party at his parents' home, a farm in Wiltshire in southern England. At the bottom of the property was a small copse with a clearing. My friend set up the sound system in a horsebox and I spent the night inside playing music and getting merry on home-distilled gin, my payment for the evening.

My friend's family were pig farmers, and his father had slaughtered one and set up a spit in the clearing, upon which the animal roasted all day. Alongside the pig was a huge silver saucepan of homemade applesauce and a pile of buns. I remember the whiteness of the pig's flesh turned to meat, rather than the pinkness of gammon and ham. The pig's body was central to the evening's events: the celebration of my friend's passage from boyhood to man's estate. And yet it was not *seeable* as a body that once lived. Gary Francione has criticized Melanie Joy's claims that speciesism is an "invisible" ideology; and there was nothing invisible about this pig, the centerpiece of the evening. To us, nothing seemed wrong with killing this animal to consume its body. It was hidden in plain sight.

I think of that pig now. I remember his physicality, the sparks of the fire and the commensality we shared as a group around his body, feeding on the meat and never according his body the same status as ours. For all my lack of awareness back then, perhaps it is something that I nonetheless remember the fascination. Although I had a vague feeling of the pig's body as an uncanny presence, of his dubious role in the rituals of manhood we were celebrating that night, only later would I ask the question, as Anat Pick does: "What would these animals have to become, and become in our eyes, to be creatures that it is forbidden to kill?" (66)

Can the body of just one animal, one pig, help us to see a whole species in a new light? To make advocates of us? When the conditions are right, when that pig is alive and charismatic, and when her companions are adept at sharing her story via social media and have a wicked sense of humor, then perhaps *yes*.

Many within the animal rights movement, especially in North America, already know the Facebook phenomenon that is Esther the Wonder Pig.[9] She was originally adopted as a "micro-pig" pet in a Toronto household by her human companions and "fathers," Steve and Derek. But she turned out not to be so mini (there is no such thing as a micro-pig). Neither Steve nor Derek were involved in animal advocacy. Yet the two men and Esther have become worldwide animal advocacy celebrities, with a quarter of a million followers on Facebook, raising over Can$400,000 to help purchase an old farm to turn into a sanctuary, Happily Ever Esther.

This phenomenon came about because of the intimate engagement and changing relationship between the bodies of two humans and that of one animal they used to consider meat. Esther was bred to be slaughtered and was somehow overlooked or rescued or escaped as a runt, of no value to the farming industry. By the time Steve and Derek realized Esther was not a micro-pig, however, they'd already fallen in love with her in the same way they love their dog and cat companions. Through this intimate proximity with an individual pig, coming to know her emotionality and responding to her intelligence, they grew cognizant that *every other pig was just like Esther.*

Steve and Derek turned vegan and created a Facebook page to share with their friends and family on what it was like to live with Esther. Their sharp, humorous commentary and Esther's clear delight in living won over hundreds of thousands of people. Because of their growing fame, and because of the new knowledge that living in proximity with a farmed animal had given Steve and Derek, they needed to go further. Subsequently, they've devoted themselves to advocacy and care for animals. What Steve and Derek have achieved, and the way they have achieved it—the speed and nature of the transition—would not have been possible without social media. However, the instigating change came at the corporeal level, in intimate bodily relation—finding, as Mary Midgley says, "emotional fellowship" between bodies.

For Donna Haraway the very specific cohabitation and co-becoming of humans and dogs over millennia have a history that explains their proximity to us and our special treatment of them. The pig, on the other hand, as Mark Essig writes in *Lesser Beasts: A Snout-to-Tail History of the Humble Pig*, has often been seen as distant and unclean. But as Aristotle once wrote, pigs are those animals most like humans. Might, as Brett Mizelle, the author of *Pig*, asks, our "disdain for pigs have to do with shame and guilt about our relationship with them?" (120)

The continual repetition of images and stories of animals shapes our thoughts and feelings about them. Social media is both enforcing and changing this . . . perhaps. For most people, even those active on social media, the chance to witness any image or message about the bodies of pigs is limited, unless it is actively sought out. Without effort, our social media forms a bubble of reinforcing likes and preferences. Yet the animal rights movement is enjoying new opportunities as the institutional gatekeeping role over what people *can* see, traditionally held by corporate media, has broken down. We are more able than ever to shape our own bubbles. This is a democratization of perception brought about and through the sharing of imagery among people. Once the door is opened just a little and we peer through the crack, we cannot unsee what we have seen, and Facebook and Google in particular, with their personalized algorithms, help us to keep seeing.

Esther's is a story of hope. Yet a billion of her brothers and sisters are trapped every year in a system of speciesist human exceptionalism that is growing worldwide. Rarely do these individuals escape. When they do, such as for our titular pig, they can change us. The ideology of speciesism is breaking down, and social media has played a major role in that. We are coming to see other images, and we see them repetitively enough to begin the change in how we relate to those nonhuman others.

The Tragedy of the Pig

What was it about the picture of the pig flinging herself from the truck that haunted me? It was not, after all, graphic. Was I using this image, as Brett Mizelle warns us against, as "a palliative exception that helps us forget that our dominant relationship with pigs involves killing and eating them" (106)? Was it in fact that, as my feed refreshed, I lost sight of her, and this was a flaw in my witnessing that exposed it for what it was: useless in the face of her fate? Or was it because I thought she was not exceptional and her freedom was only momentary? Did the truck driver see the pig from the wing mirror as she jumped? Did he stop, scoop her back up? Was she already rendered into chops?

The truck driver didn't notice his lightened load, in fact. "If we hadn't have got to her, she would have undoubtedly found herself at the end of a blade wielded by a local who would have enjoyed pork for days to come," a police spokesperson stated, taking credit for rescuing this pig and also clarifying her gender.[10] She, our pig, found her way from the hurrying cars to either a kind soul, or a wide open space where she could hide, until she was picked up by the police.

"She saw one chance of freedom by clambering on the backs of the others and took it. She deserves her chance of life and she has got it," the spokesperson carried on. "She will never be eaten here." The pig asserted her wish for freedom and so jumped into personhood, as did the cow who escaped from the holding pen of a slaughterhouse in Omaha, Nebraska, with whose story Timothy Pachirat begins *Every Twelve Seconds*, his book on "industrialized slaughter and the politics of sight." The cow finds her way down an alley that leads to another slaughterhouse and is shot by the police—right in front of the slaughterhouse workers on their break. They are "livid with indignation" (2) at this treatment of the animal, who has, for this moment, broken out of the mass. Pachirat writes: "[T]he physical escape of the cattle from the Omaha slaughterhouse is also a conceptual escape, a rupture of categories. . . . Conceptually dangerous, their escape threatened to surface power relations that work precisely through confinement, segregation, and invisibility" (4–5).

The differences with the human animal are writ large. We do not need to do anything to deserve our chance of life; for nonhuman animals there is no automatic right. Sometimes, however, a chance arises. That chance is a small loophole or the eye of a needle through which the nonhuman animal must pass if she is to be permitted the privilege of living. If the animal sees it and takes it, and the external circumstances (sympathetic humans, food, careful drivers) are aligned, then the animal may not only survive the risky escape attempt but also be granted her "chance"—a chance conferred on her by the most powerful group of humans to come into contact with her story.

For our pig, this group of humans was the local police, who avowed that *she would never be eaten here* because she expressed her desire for life. What this implies, of course, if we are to let surface those power relations that such an escape threatens, is that we have chosen to agree that all the other pigs in the truck did not express a desire for life; that they did not take their "chance." And yet we know that if they did, this unique story would have become too much of a shock for the system. The police would've had no resources to adopt fifty pigs. They would have all been rounded up and put back on the truck and taken to the slaughterhouse.

What haunts me here is what goes unseen: what she, our pig, was escaping from. The slaughterhouse. The frightening conditions inside the transport truck, and the life she'd lived up to that point. Perhaps what haunts me most is that in my life up to this point I'd done so little to listen for the claims to life and freedom of all the animals who had gone before. I remained silent and anonymous.

The philosopher Jacques Derrida turned at the end of his life with great commitment toward animals (in plural and also the particular animal that was his companion, a cat). Derrida brought forth the idea of not only seeing, but being seen; in fact, being "seen seen" (13). This concept is at the very heart of the ways in which we treat nonhuman others. For Derrida, to be seen by the other and also to accept that we are "seen seen"—that the nonhuman animal knows that we know we have been seen by her—is at the heart of how change in our thinking about animals will come about. The Kantian, rationalist view of other animals is based on the denial of being seen by them; and our dominant ethical relations to nonhumans are based on this ability to see them but "without being seen seen naked by someone who, from deep within a life called animal, and not only by means of a gaze, would have obliged them to recognize, at the moment of address, that this was their affair, their lookout" (14). For Derrida, the image of the suffering animal is pathetic. But these "pathetic" images "open the immense question of pathos and the pathological, precisely, that is, of suffering, pity, and compassion" (26).

Through these images, I bear witness. But have I done more for the animal than the philosophers whom Derrida attacks? Those who have "no doubt seen, observed, analyzed, reflected on the animal" but who "neither wanted nor had the capacity to draw any systematic consequences from the fact that an animal, could, facing them, look at them, clothed or naked, and in a word, without a word, address them" (13). These philosophers, says Derrida, "have never been *seen seen* by the animal" (13). They have "taken no account of the fact that what they call 'animal' could *look at* them, and *address* them from down there, from a wholly other origin" (13) [all emphases in original].

Through images shared around social media I came to know that the animal could address the witness. I came to understand my brain–body–world entanglements in which the animal was already present. I was aware of how imagery, videos, and stories and their distribution through the rhizomatic connections of social media were providing some of the most influential tools available to animal activists. But although I was bearing witness to another's experience of life and body, I also understood that seeing such things through social media was not enough. I watched again the videos of activists bearing witness at vigils outside slaughterhouses and holding dead bodies in silent testimony to their lives. Here was the difference: these were activists who bore witness but who were also being "seen seen." My witnessing was painful, but I was not myself seen, physically and in my body, by others. I had not looked into the eyes of a pig or been addressed. Witnessing is an act, a phenomenological experience performed through the body. I knew that if I wanted to be an active witness, and if this was going to fundamentally change my relationship with other species, I couldn't do that from behind a screen.

Our pig's fate is to live out her life as a mascot for the local police—not surrounded by other pigs with grass under her feet but in a pen in a concrete yard usually inhabited by a police dog (and what has happened to the dog, in a country where in some parts dog meat is a delicacy?). To be sure, it's a better life than the one she was fated for: a terrifying last few moments, then the stunner, the sticker, and the slaughter. There she would have disappeared into the air proper, become invisible with her forty-nine other brothers and sisters crammed into that truck.

I am pleased for her. The police have named her Babe, one guesses after the eponymous movie character. Then I look again. I see that other pig in the picture, the one staring back at us still in the truck, less brave, perhaps, or more confused, stuck under the netting of barbed wire. It is *her* face, I understand now, as well as Babe's escape, that haunts me, and why this picture has stayed with me. That is: *look closer.* In her eyes, barely visible, I see she recognizes her fate. She knows where she is going. And, perhaps in a realm beyond the immediately graspable, I feel that she knows I know. In her eyes, I and my species are seen. What could I do to save *her*?

As I've suggested, images of nonhuman animals shared via social media have been an essential factor in my becoming an animal advocate. But bearing witness to images as they are circulated on social media is not enough to bring about the wholesale change that abolitionists and liberationists call for. It is not only a "politics of sight" that is at the heart of animal advocacy work but rather a politics of the body and the senses. To exclude the sensing body from advocacy work reduces its effectiveness. As Buller, quoting the Canadian scholar Stephanie Springgay, puts it: "[F]arm animals offer us the potential or the promise of an affective mattering derived from 'a more proximal, contingent and bodily form of thought'" and this is made up of what is mostly lacking from images: "noises, smells, movement and shared vitality" (156). Changing our relation to other species comes through being in company with them, body to body. This is especially true of those nonhuman bodies whom we disappear into the system. This "bodily form of thought" is critical for changing our species' relation to others. To do this, like Derek and Steve with Esther, we need to come closer to those others, to understand how they are treated and advocate for change.

Fifteen years ago I believed moving to the Middle East would take me to the root of the world's most pressing problems. Now I understand that I've arrived at that point after all. That point is reached by coming closer to the body of Babe, and closer to all the pigs, within an industry that is spewing forth greenhouse gas emissions in ever greater amounts. It is time to pay a little more attention to the other element of this story, the thin air around our bodies, the air that we all breathe. Because there is no greater threat to our collective future than climate change and the greenhouse gas emissions that intensive animal agriculture produces.

3

Identity Change, Climate Change

WE HAVE NOW come, as Tolstoy asks us to,[11] closer to the suffering of the other, to the physical bodies of nonhuman animals. By practicing the moral attention that Lori Gruen calls for in attending to the bodily needs and desires of nonhumans, rather than to their "rights" (which are constantly denied anyway), we can understand more fully what might constitute our brain–body–world entanglements with these living beings.

This task is overdue. Historically, little study of society has been conducted from the point of view of the animal's body or of animals' bodily importance for our shared experience. The main exceptions to this are found in feminist care theory: in the writings of ethicists such as Carol J. Adams and Greta Gaard, and within some areas of phenomenology, sociology, and cultural studies. The "animal turn" in the humanities has offered new concepts and ideas. But there is much more work to do in understanding what the controls, restraints, body projects, body option, and body regimes that we place upon nonhuman others might mean for them and us, enmeshed as we are. This work needs also to take into account identity processes; not only our identities, as philosopher Matthew Calarco explores, in continuity or in distinction with nonhumans, but about how, together, we can remake ourselves in response to the changing, threatened environment in which we live.

We've met Babe and Esther. What about the others?

The Pig in Production

Pigs are the most eaten land animal. Although more chickens are killed each year for food, in terms of the amount of meat product consumed, pigs top the list. (In terms of total numbers, marine animals are far and away the most exploited.) On Earth at any one time, around a billion pigs are being processed to deliver over one hundred million metric tons of meat each year. The number of pigs is growing, mainly because of increased affluence in China, other parts of Asia, and South America. Like our Babe, half of the world's pigs are found in China, which produces around fifty million metric tons of pig product for consumption. China's pig farming is still relatively traditional, with grazing and mixed means of processing through small farms. But the country is moving to a more industrialized system. In the next thirty years, as the world becomes more populous, meat consumption is likely to double. Much of that will be made from pigs.

Whereas welfare laws have meant that pigs in Europe are spared some of the worst abuses, globally they still undergo many horrific practices that are considered "industry standard" and are therefore legal. This is especially the case on American industrialized farms, which are increasingly the model for pig production facilities globally. Unwanted piglets are taken by a back leg and slammed to death on the concrete floor. At three to five days old male piglets are castrated without anesthetic to reduce "boar taint," a disflavoring of the meat that occurs when pigs are ready to mate due to the release of the hormone androsterone. One way around this violence would be to kill pigs at an earlier age, but then they would produce less meat.

Then there's the gestation crate, which is a metal cage used to pin the sow into place when she is pregnant and weaning so she is not an economic threat to the new produce—her babies—by rolling over and crushing them. (This phenomenon almost never happens when pigs have room to create their own nests.) The crates are arranged in long rows and the sows are first forcibly impregnated at around seven months of age. The mothers live out their lives in a cycle of pregnancy, birth, and nursing until they are worn out and no longer profitable, at which point they are sent to slaughter.

These sows—social, intelligent, active animals—suffer porcine stress syndrome (PSS), their version of PTSD. Undercover investigators have reported pigs who have escaped by attempting to undo the locks of fellow pigs' crates, as the naturalists Gilbert White and Cicely Howell observed them doing to the gates of fields in the eighteenth century. (Unlocking the crates of her fellow incarcerated creatures is one of the first acts of Pig 323 in Gene Stone's novel *The Awareness*, which tells the story of nonhuman life coming to consciousness.)

Recent public pressure has led companies such as Walmart, McDonald's, Burger King, and others to ask their pork providers to phase out these crates. However, they are still employed in the majority of pork facilities in the United States, where around six million pigs are used for breeding. Even Temple Grandin, professor of livestock behavior and welfare at Colorado State University, autism activist, and consultant to the "livestock" industry on animal behavior who claims to "think like a cow" and designs the animal-handling equipment to take those animals to slaughter, knows it is wrong: "Confining an animal for most of its life in a box in which it is not able to turn around does not provide a decent life" (27). Even though their banning would result in only a little less overall suffering of all farmed animals, gestation crates are so "cruel and profoundly unnatural" (135) as Gene Baur, cofounder of Farm Sanctuary, says, that while ending exploitation must be the goal, making gestation crates illegal would be a welcome victory along that path. On my own journey toward animal activism, a picture that upset me more than any other image of factory farming consisted of long rows of pigs in gestation crates, like the human battery farms that Neo awakes to in *The Matrix*.

There are other abuses to which we subject pigs. We dock their tails and file their teeth to stop them from gnawing at each other in their distress. Even then, pigs will sometimes cannibalize their wounded, sick, or dying conspecifics if the "downed" are left to fend for themselves. And then, of course, we slaughter them, long before the end of their natural lifespan. The abuses documented by undercover investigators include men raping sows by shoving iron bars into their reproductive tracts; punching, kicking, and burning animals; using electric prods excessively to keep kill lines going; and more. Some of these practices are illegal. As I scoured the pages of books, websites, and social media in my journey toward understanding the experience of pigs' lives with which I was already entangled, I realized that the pig in the industrial food system does not live but merely exists in pain, until she either escapes or is violently put out of her misery. I imagine it was this misery that she already knew rather than the threat of the slaughterhouse that compelled Babe to escape.

The industrial system is not only a horror story for the pigs. Animal agriculture, the world's largest industrial complex, is impacting our environment in catastrophic ways, and its effects upon climate change are becoming clearer. It is ever more critical (even for those of us not yet on "the animal side") that we make greater attempts to understand how nonhuman animals within our agricultural industries are already entangled with our own experiences. Our survival may well depend on such enlivened, embodied understanding, and our excuses not to act, like the air we breathe, are wearing thin.

Coming to Care about Climate Change

The image of Babe arrived in my social media stream at a time when I was thinking about the relationship between we animals and air, or, more precisely, climate—our atmosphere that, if we scooped it all up into one pot, has the same volume, give or take, as the Mediterranean Sea. The life-giving air we breathe is a thin slip of vapor into which we pump upward of sixty billion metric tons of greenhouse gas emissions each year. Those summery clouds in the background of the picture of Babe belie the great threat of climate change, as well as the carbon-laden smog that affects so many Chinese cities. In many ways we are trapped in a juggernaut about to crash, and there doesn't seem to be anywhere *we* can jump.

I've been studying climate change for a decade and a half. I'm not a scientist, but rather I have examined the cultural narratives of our responses—the *story* of climate change. First, I explored, while working at the media charity, how people were acting on climate change through development projects. Then, as a scholar, I analyzed the media and literary representations of the crisis, looking at how we visualize it, imagine it, and what we say about ourselves as affected (or not) by the threat.

What I've seen is much of what others see: apathy that hides anger and frustration, helplessness and denial; disbelief, ignorance, and often mere forgetting. These are all examples of what Naomi Klein calls our "great turning away" from the matter. Klein gets it right. We don't come closer to climate change because we're scared that if we do it changes everything about our global system; that we'll lose all of the comforts, privileges, and luxuries brought to us by economic growth. Klein's is a systemic view: deregulated capitalism is "at war" with the planet and it has to change. My work has been to try to understand what will turn us toward the difficulty of change and how narrative helps us do that.

I myself am paradigmatic of the problem: in 2007 I turned away, too. A year into a research program, I became depressed. Having read so much in such a short time about climate change, I was not constitutionally strong enough to deal with the end of our recognizable world. I dropped out of academia and returned to a job that didn't deal with the subject. Yet the issue was always there, nagging at me to do more. Part of my turning toward the crisis has come because of the deeper "coming to care" of my move toward animal advocacy. Working on climate change is now not just for myself or my species, but for every animal. But 2007 was not where my journey into the crisis began. For that, we need once more to go a little further back.

—

I first became aware of climate change in 2002. I was employed at the time as an editorial manager at a media charity. Our organization was small but operated globally, involved in the new world of connected media, the Internet, and mobile phones, and the benefits they could bring to poorer communities. We were working with a technology team in Norway and I was sent to help manage the project. I traveled back and forth a dozen times to Skien, a small town southwest of Oslo and the home of Norway's most famous playwright, Henrik Ibsen.

One of my early ambitions in life was to swim in a Norwegian fjord, a feat I finally achieved on a hot July day near the town of Brevik. That year had already set new records for the warmest January and first quarter on record. Despite the summer heat, the fjord was freezing and I soon had to get out, dry myself off, and wonder if it was worth it after all.

Perhaps it was. Not long after my dip, on the coastal journey from Skien to the airport I paid more attention than usual to the hillsides of firs—vast, glorious stretches of forest above the eyeline. It struck me that such boreal forests—mainly Norway spruce, Scots pine, and birch—were breathing. It came to me with a newly felt reality. Perhaps it was the seawater in my sinuses, but I understood that each time I took off in a plane I was polluting their air with emissions, some of which they could breathe out as oxygen, but increasingly more of which they could not.

The Norway project was to develop a new management system for our network of websites, linking the international offices to coordinate content across important themes, including climate change. Since the turn of the millennium, climate change had flooded popular consciousness via the media, emphasizing it as an urgent, global environmental crisis. This was the time of climate activist Mark Lynas's *High Tide*, many years after Bill McKibben's *The End of Nature* (published in 1989) and Al Gore's *Earth in the Balance* (1992), but before *An Inconvenient Truth* (2007), which finally brought mainstream attention to climate change.

That day on the bus from the home of Ibsen to the Oslo airport, I finally connected my individual behavior with the global picture. Here I was, jet-setting (on budget airlines) between European and African hubs to build an electronic media center to help share news on climate change. Yet I was heavily contributing to the carbon footprint of the human population without the knowledge or wisdom to do better.

Back in the 2000s, air travel was considered the worst of all possible individual contributors to climate change.

Airports were the target of direct action from organizations such as Plane Stupid, whose activists chained themselves to fences on the runways.[12] The biggest single thing people could do, we were told, was to stop flying long-haul if we could, and absolutely stop flying short-haul. This is still the activity that many who want to act on climate change feel most guilty about.

Over the next few years, climate change became *the* issue in addressing our global environmental challenge. It is now almost universally accepted that anthropogenic climate change is likely to see global temperatures rise by more than two degrees Celsius in the next hundred years due to the increased concentration of greenhouse gas emissions trapped in our atmosphere. These greenhouse gases, most notably carbon dioxide and methane, are the product of our global capitalist system. These rising temperatures take us toward what are known, perhaps infamously, as tipping points, beyond which things will rapidly and chaotically speed up without hope of reversal. The melting of the Greenland ice sheets and the rising of our oceans will lead to extinction of flora and fauna, mass migrations from countries subsumed by water or grown too arid, and the collapse of ecosystems. If we do not rapidly reduce our fossil fuel emissions to around ninety percent of current levels, and speed up practices to mitigate and adapt to the damage that is already irreversible, then the global community will experience orderly or disorderly decline into radically altered social relations. The world will not exist as we understand it socially or culturally, and certainly not environmentally. This is the "collapse" scenario that the likes of financier George Soros and biologist E. O. Wilson now entertain, and that those working in climate change science have been preparing for already.

My revelation in Norway was the moment I came to bear witness to the impacts of climate change. The myriad voices collected by first the World Wide Fund For Nature (WWF)[13] and now the Christian Reformed Church's[14] Climate Witness Project are not only individual people with stories but metaphors for the need to make climate change visible. To bear witness is to see the truth of what happens during an event. For an event of such magnitude as climate change, none of us are bystanders or outsiders.

The Role of Animal Agriculture

In 1971, Frances Moore Lappé's *Diet for a Small Planet* first connected the environmental impact of meat production with people's diets. Although the book has sold over three million copies and led to the establishment of the California-based Institute for Food and Development Policy (Food First), by the time of my Norway revelation in 2002 very little was being said about the impact of animal agriculture on the climate. Eyes remained on the fallout from 9/11 and the impending war in Iraq. It would take another four years for the Food and Agriculture Organization of the United Nations (UNFAO) and their report *Livestock's Long Shadow* to link climate change and animal agriculture. (That was the same year, in fact, in which Anna Lappé, Frances's daughter, wrote and published *Diet for a Hot Planet: The Climate Crisis at the End of Your Fork.*)

Livestock's Long Shadow suggested that animal agriculture was responsible for around fourteen to eighteen percent of all greenhouse gas emissions. The report added food to the list of industrial sectors and everyday habits that were visibly responsible for climate change. Before that, as Elaine Graham-Leigh argues in her book *A Diet of Austerity: Class, Food and Climate Change*, official analysis had not included what we ate. As she attended the Campaign against Climate Change demonstration in London in 2008, Graham-Leigh noticed a shift in the messages on the placards and attendance by organizations such as The Vegan Society.

Then Chair of the UN Intergovernmental Panel on Climate Change (IPCC), Dr. Rajendra Pachauri, teamed up with singer Paul McCartney to argue for a reduction in meat consumption. PETA quoted the report in a challenge to Al Gore in March 2007, asking him to go vegan, and it was the main scientific source in Dutch activist and politician Marianne Thieme's documentary film *Meat the Truth* (2007). But the issue as a political question did not stick, lost in the overwhelming complications of international treaty negotiations, sensitivities around dictating people's food choices, and (perhaps most of all) the global financial collapse of 2008.

The Worldwatch Institute reprised the issue in 2009, when Robert Goodland and Jeff Anhang claimed that the true figure for animal agriculture's contribution to greenhouse gas emissions was fifty-one percent. Yet while government strategy reports and international papers referred to the question of meat and dairy consumption linked to climate change, no government seriously took action on telling people to eat less meat. (In 2015, the United States' dietary advisory council finally began to argue that guidelines should also take into account environmental impact, to vehement resistance from the meat and dairy industries.)[15] Some may have held placards at demonstrations raising the climate costs of meat consumption, but the major environmental organizations—Greenpeace, WWF, Friends of the Earth—did not change their focus from fossil fuel campaigns to meat *production*. Countries of the global North were battling to save their economies and the environment could wait. But for how long?

In 2013, the UNFAO revised their figure of greenhouse gas emissions directly from livestock to 14.5 percent—which is still greater than for all global transport combined, and excludes the transport required to move animal food products around the planet.[16] Other current estimates put animal agriculture as responsible for as much as twenty percent of total global greenhouse gas emissions as part of the umbrella agriculture, forestry, and land use sector.[17] Whatever the true scope of the figure between fourteen and fifty-one percent, if roughly one fifth of all your bodily ills were coming from a single bad habit, wouldn't you do something about it?

But what ails the body isn't always something we can rid ourselves of so easily. Habits are hard to break; to stop consuming animal products seems the hardest of all. Yet evidence points to the fact that we need to move toward a plant-based diet if we want to ensure a stable environmental future. According to the economist Richard Oppenlander, eating animal products creates ten times more fossil fuel emissions per calorie than plant-based foods. Nonetheless, for example, nearly ninety percent of soy grown globally is used as feed for nonhuman animals in agriculture, while overall about half of the planet's crops are fed to nonhumans.

Some may consider this a crime against humanity when over a billion people go hungry every day. It is also massively inefficient. When we feed grains to nonhuman animals, we lose up to ninety percent of the protein, ninety-eight percent of the calories, and one hundred percent of the carbohydrates. Nonhuman animals need to eat huge amounts of food every day to produce just one body that will be rendered into meat products for we humans and our companion animals to consume. Some scientists are attempting to "fit the animal to the system" by, for example, breeding chickens without feathers so they don't waste calories growing "unessential" body parts.[18] But this is playing (God) in the margins.

Animal agriculture is the largest user of fresh water and the largest single reason for rainforest destruction (to graze cattle and grow soy and grains to feed farmed animals). It is also one of the globe's major polluters of land and water tables, and creator of sea "dead zones"—areas of the ocean so polluted by animal agriculture and fertilizer runoff that nothing grows there. Other books argue the point with more data, depth, and economic analysis than I can include here, such as Lisa Kemmerer's *Eating Earth*, David Robinson Simon's *Meatonomics*, Richard Oppenlander's *Comfortably Unaware*, and Philip Lymbery and Isabel Oakeshott's *Farmageddon*. Some environmental groups are now catching up with their animal rights kin and programming climate activism that engages with this knowledge. Governments are slower to act, and most, along with some corporations, are still moving *toward* the catastrophe, and indeed bringing it closer. This is occurring at a time when, apart from a vocal and powerful segment of right-wing Euro-American society, a consensus exists that humans are causing climate change and that it will *change everything*.

Climate Veganism

So what are we doing about it? Is the world ready for a plant-based diet?

Forty-five years ago, Frances Moore Lappé advocated an "ethical vegetarianism" as the answer. Today, an approach is taking shape in cities around the world under a "climate vegan" banner, remaking the link between animal agriculture and climate change. At present, these movements are led mainly by animal activists and advocacy organizations. As the film *Cowspiracy* documented in 2014, large environmental groups such as Greenpeace and the Sierra Club have been largely silent on the matter; the film argues they are too concerned with upsetting their funding base to promote messages to eschew meat and dairy.

Support for this "climate vegan" approach is not universal. Many have pointed out that a shift to a plant-based diet is a generational movement. Yet climate change seems not to allow time for generational change—or for activists to keep on using advocacy similar to that which has been used for many decades, during which time global production of meat products has increased fourfold, twice as fast as the human population.[19] And perhaps there still remains a strong enough business case for change, which is slowed or even obstructed by radical activism. As the authors of *Chomping Climate Change* argue, a stronger business case must be made to replace animal-based foods with plant-based foods before it's too late.[20]

The writer and campaigner Elaine Graham-Leigh argues that the focus on consumer choice benefits the corporate producers of food products, who are allowed to continue production without challenge or with only minor tweaks (e.g., phasing out gestation crates does nothing to reduce emissions). She suggests this approach lays the blame for climate change *not* at the door of rich and powerful elites, but on those overindulgent over-consumers of processed meat foods: the (obese) poor. Graham-Leigh wants us to focus instead on the producers and the wastefulness that is not a by-product of a capitalist system but an *essential multiplier*, creating surplus value (profit) by adding stages into the production process. It is not the natural constituency of the food product itself, she argues, that is the problem—such as cows being "naturally" high emitters of greenhouse gases. Rather, she says, the problem is the process. A focus on moving to a plant-based diet does nothing to change a capitalist system that seeks profit by maximizing waste potential. It will just fly *more* vegetables around the world.

Graham-Leigh's argument is with the U.K. green movement rather than the global or North American animal rights movement. She feels the U.K. greens have co-opted the "no meat" argument without thinking through the causes of climate change. But for a book that asks questions about the relationship between food, consuming bodies, and environmental catastrophe, a lack of interest in the way that capitalism is built upon the exploited labor of human *and* animal bodies seems curious for a self-described Marxist. It also misses the point, made by ethicists such as Carol J. Adams, that animal exploitation is such a central part of the system that to end animal suffering would break that system. A new alliance between animal activists and the Marxist Left is emerging, in academic scholarship at least, around this question of the exploited, laboring bodies of both humans and nonhumans within the destructive capitalist system, which she avoids. Graham-Leigh also ignores plant-based food companies such as Beyond Meat and Hampton Creek Foods, which have made the connection between animal agricultural *production* and climate change, and have attracted funding from capitalists such as Bill Gates to produce their meat and egg alternatives.

Looking at the "climate vegan" approach in terms of the exploited bodies of laboring animals, it is easy to understand why campaigns to highlight how intensive animal agriculture impacts the world have focused on the cow's body. Campaigners emphasize the high levels of enteric fermentation leading to methane emissions and the land destruction and water use to raise and graze cattle. Whereas it is estimated that, on average, showers taken by each of us consume 5,100 liters of water per year, producing just one pound of beef uses 5,200 liters, which is also two and a quarter times the water used for pig meat production. Conscientious campaigns to urge individuals to reduce their carbon footprint by taking fewer showers or changing lightbulbs are derisory if they do not also focus on individual meat consumption. The answer, however, can't be to switch from eating cows to pigs. After all, a vegan diet produces *thirty times* fewer greenhouse gas emissions than an average meat eater's diet.

So let's not forget Babe. The pig is an exemplary animal as an advocate for change, not least because the meats produced from pigs' bodies are the most consumed in the world. Any focus on cattle threatens to circumvent emphasis on the root cause of our ethical and environmental crises. And the pig has her own symbolic power. She regularly appears in our Western narrative history, in our stories and books and films, and we are used to entertaining the image of the pig as a creature that helps transgress and blow apart species boundaries. She is a valuable ally in this story, a body with meaning for us not only when at ninety degrees dead or jumping into life.

Climate Veganism, Identity, and What Works

If we already know that animal agriculture is the largest cause of greenhouse gas emissions, why don't we just stop eating animals? How come we've not *changed everything*? Is it to do with the message?

Analysts Susanne Moser and Lisa Dilling suggest that because of the lack of direct experience of climate change, the issue becomes fundamentally one of communicating the experience through imagery and illustration. For communications scholar Birgitta Höijer, climate change has become known to us visually, and this has been done through the "communicative processes by which a new phenomenon is attached to well-known positive or negative emotions, for example fear or hope. In this way the unknown becomes recognizable as, for example, a threat, a danger, or something nice and pleasurable" (3).

Climate change has been communicated to us in dramatic, crisis-driven narratives, with imagery of generally negative outcomes—flooding, forest fires, melting ice—and messaging focused on impacts rather than solutions. But as Saffron O'Neill and Sophie Nicholson-Cole have nicely summarized for us, "Fear won't do it." Such messaging has not brought about the desired change. Instead, scholars have attempted to influence environmental communication specialists to frame their campaigns around solutions. Nor is the answer to why we haven't responded adequately to climate change to do with the amount we know. Social scientist Dan Kahan at Yale Law School has proven that it is not an "information deficit" that is the cause of inaction. For Kahan, a more reasonable analysis is that "respondents predisposed by their values to dismiss climate change evidence became more dismissive, and those predisposed by their values to credit such evidence were more concerned, as science literacy and numeracy increased" (742).

That is, the answer doesn't have to do with our taste preferences or how well we understand the science. The answer has to do with who we feel we are and how we form our identities in relation to our social groups, such as my feelings toward my grandfather. For example, the

Climate Outreach Information Network, based in the United Kingdom, focuses on publications and training for environmental communicators around identities—such as political and national identities, young people, and people of faith.[21] They produce materials on "how to talk climate change with the center-right." They have not yet, however, produced a report on "how to talk climate change with meat eaters." But for most living in Western—and increasingly developing—societies, one of the central pillars of identity is as prepackaged meat eaters. To consume animals is, for the majority of the human population, *normal*. To do otherwise is considered abnormal and to be vegan or advocate for animals is often, therefore, deviant, perhaps even *pig-headed*.

So the answer to this question—*How do we change everything?*—may be that we first need to change our identities as the humans who have caused the problem. Over the last decade, a community of scholars and activists involved in turning the world toward climate change has named identity as the "missing link" in making climate campaigns effective. A move toward an identity-based approach to climate change has emerged, focusing on the part played by our individual, social, and group membership identities in responding (or not) to the threat. What I and others have been researching is this same question Klein asks—*How do we change the system?*—but from the point of view of identity: *Who are we as agents of change in the system?*

Much of the research has developed out of Identity Process Theory. As social scientists Rusi Jaspal, Brigitte Nerlich, and Marco Cinnirella put it: "[I]ndividuals need to maintain appropriate levels of particular identity principles and they will behave in ways that restore appropriate levels of particular identity principles when they are threatened" (122). These principles include continuity, distinctiveness, self-efficacy, and self-esteem. When we feel as if our identity in any of these areas is threatened through coercion toward behavioral change (say, foregoing meat) we rush to defend the principle, rather than respond to the threat (animal agriculture's impact on the planet). Most animal advocates will understand this reaction to their work, and understand it viscerally. It is also why many messages about climate change—loaded with shock-, fear-, and shame-inducing communications—have not worked. Jaspal and his colleagues suggest that it is necessary "to examine the ways in which particular behaviors (e.g. use of one's car; the consumption of meat) might impinge upon identity processes" (118). The more a change is seen as threatening for identity, the less willing we will be to endorse it. Behavioral change that interferes with daily life convenience—such as what one eats—is particularly threatening.

If we know that identity is constituted through our body and bodily relations to the other—and the body politic as a whole—what does this mean for tackling the threats people feel when faced with the crisis of climate change, particularly when they are asked to give up eating meat?

As a society, at the local or national or global level our relation to climate change is constituted by this fellowship and our social group belonging, which is formed at the insistence of our bodies. What Naomi Klein argues for—that we must give up our privileges of economic growth (which includes, I would add but which Klein avoids naming, the meat and dairy and other animal products we use)—will be a huge change in how we structure society. We know that social worlds are structured by our bodies. So to respond to climate change begins with the body, too. To do this, animal liberation, vegan, and climate activists will need to work together to change people's behavior through changing their bodily practices: to remove negative repetitions, reinforce positive behaviors, and allow space in society for vulnerability and shame. It may turn out to be a slow process, but it is fundamental to lasting change.

In this existential effort, pigs like Babe and Esther have a central role to play. What is happening when we see a pig like Babe escape? To "think through the body" about such escapees, we might say these pigs (and cows, and others) are changing their relationship to us through the movement of their body. Animal advocates and slaughterhouse workers all testify that each animal knows that the slaughterhouse is a place it doesn't want to be. When they do escape, we no longer look on them as things to eat because of their *movement* toward us, away from the invisible spaces we have assigned them. We are forced to see them differently. They have changed their relationship as the member of one species to the members of another—us—by forcing us to recognize them as creatures with a desire for life. Through their bodily motion.

Babe and the other pigs in the back of the truck were suffering the traumas of captivity, followed by the disorienting journey to slaughter. Maybe they'd been in gestation crates and this was their first sense of being able to move their bodies. Babe *could* run, and *did*. She mobilized her body, choosing autonomy over destiny.

The effectiveness of any response to both climate change and animal rights will depend on having radically different senses of who we are and what it is permissible for us to move toward. This will involve changing people's identity principles and their social group belongings. Such work cannot focus only on individual human identities but also on the identities of those with whom we are entangled. To take just one individual, Babe's identity has become an element in how change might occur in my body. Or to put it another way: Who *Babe* is shapes who *I* am.

This is the less traumatic version. Because who that *other* pig was, the one looking back at us from out of the truck, who did not jump, also shapes who we are. To see and feel the other in this way is more likely to lead us toward a climate-friendly plant-based diet, and to bring us back into a "balanced and clear self-concept" of who we are in relation to a world as vulnerable and material as we have finally admitted ourselves to be. For myself, becoming an animal advocate has been about this reappraisal of my bodily sense as a corporeal being living alongside others, all dependent on this singular environment, Earth. Compassion for others means sustainability for ourselves.

And so we move on to Part II, where the question becomes: *Now that I am an advocate for animals, what's the most effective I can be?* Having done the groundwork of experiencing brain–body–world entanglements, we begin the physical journey of the trip I took in the summer of 2014, where I would learn what it means to be an embodied being, a human animal on the front line of animal advocacy.

Part II

How Can I Be a Good Animal Advocate?

4

A Narrative Animal Advocate

MY WINSTON CHURCHILL Fellowship provided me with the support to travel across North America for two months to learn from writers, educators, and advocates for animal protection. The award originally came out of marine education work with whale charities in the United Kingdom, so the plan was to begin at the Vancouver Aquarium (VanAqua) as a visiting writer-in-residence, before heading to the San Juan Islands to work with cetacean protection organizations. Between the award of the fellowship and starting the trip, two things happened. First, VanAqua became the target of an animal rights campaign, including from primatologist Jane Goodall, to end their captive whale program.[22] The second was that, although I was still new to all of this, my vegan outlook was rapidly changing into one of animal liberation. I came to ask, as Sara Ruddick compels us to, not what rights nonhumans have, but, more directly of the animals themselves, "What are you going through?" (348). Consequently, I understood I could not work with organizations who captured and bred other species for the purposes of entertainment.

I, therefore, changed the focus for my time in Vancouver. The picture of Babe in her truck on her way to slaughter had instilled a need in me to bear witness to the banal repetition of this last journey made by billions of nonhumans. The morning after I arrived, I put my running gear on and jogged over to join activist group Liberation BC at their weekly vigil. My first ever work on the front line of activism on a wet Friday morning was not to bear witness to the bodies of pigs, however, but rather their smaller and more numerous brethren at Hallmark Poultry Processing, a chicken slaughterhouse on the corner of Commercial Drive and East Hastings Street.

The Vulnerability of Being Seen

"If you smile and wave, it's an invitation," explained Mary-chris Staples, organizer of the regular Friday vigil, alone that morning until I arrived. She stood outside the slaughterhouse, a nondescript block hidden by broad-leaved trees. A placard hung around her neck with an image of a chicken and the story of what happens inside the building. A dozen more placards were tied to lampposts and electricity boxes. "People can see you're not here to shout at them. They can see that, 'Look, here's a happy person doing something for the animals,' and that's the invitation."

I stood with Mary-chris as she smiled and waved at the passing traffic. I was relieved as much as anything that the protest was focused on this friendly invite, but I couldn't quite muster a smile myself. I felt embarrassed and anxious. Although my body was at the protest, a part of my identity was still with the passersby. But this was, after all, only my first try. While we talked, Mary-chris kept smiling and waving. Plenty of people smiled and waved back, accompanied by a sizeable orchestra of honking horns. Mary-chris told me about one young boy with his parents whose car had stopped at the lights. He read the placard and then his eyes lifted to the building. She could see something happen in his mind.

"Even if it's just one person . . . ," Mary-chris added.

And then it was my turn.

"Can you hold this while I run to the washroom?" she asked, taking off her placard and handing it to me. She ran up the street, leaving me on my own.

I suppose it carries some symbolic weight that as someone who has hidden for most of his life behind books, screens, and the written word, I took my first small step across the line into animal activism by hanging a placard around my neck. I wore it like Nathaniel Hawthorne's scarlet letter, or one of the condemned in a photo taken before the show trial and summary execution. I knew that I was now *seen* as an activist, finally exposed. I was too self-conscious to wave. But I stood there and I faced the traffic. No one honked. I wasn't smiling, so no one smiled back.

Then a bus driver who passed by every day raised his hand in a friendly manner to me. And I thought: *Yes, okay*.

When Mary-chris returned she thanked me for doing my bit.

What did I do? I just stood there. But I *was* a body in invitation. I was a human choosing to move my body in a way that acted as a solicitation to think about the bodies of nonhumans without the freedom to move. Knowing how we embody trauma, sociality, and experience, this bodily compassion—this *enact*ivism—is more potent than we realize. For those who are used to demonstrations, protests, and rescues, perhaps such a physical expression is taken for granted. For a person making his first contribution, the discomfort was a lesson in the ways that, as Chris Shilling has said, we develop our direction and purpose "on the basis of the practical engagements" we have with our environment "as a result of the situatedness of embodied existence" (245).

The slaughterhouse facilities—a warehouse for the live arrivals, and the processing areas on the other side—were split by a back alleyway. This alleyway was a public road and so we were free to wander down it. It was here that Mary-chris showed me the workings of killing. On one side, chickens were stuffed, twenty to a crate, and those crates piled thirty high. A fast-moving forklift truck driver in dark blue overalls and a thick white mask over his face then moved two stacks of five crates each. The forklift brought forty crates from the towering columns to the open door of the warehouse, and from there the driver steadily loaded the two by fives onto an automated conveyor belt at the open end of the slaughterhouse. The chickens were minutes from death. They were forty-five days old.

"Just babies," said Mary-chris. She held up two fingers as a V-sign to the birds in the crates. "Bye, babies," she said as we took some pictures. The thought flashed through my mind that the ultimate sponsor of my trip, Winston Churchill, also used and became famous for this V-for-victory sign, long before victory was ever in sight. (Curiously enough, in a 1931 essay for *The Strand Magazine* Churchill also predicted synthetically grown meat: "We shall escape the absurdity of growing a whole chicken in order to eat the breast or wing, by growing these parts separately under a suitable medium.")

The door adjacent to where we'd been standing was in effect the garbage chute. The roller door was open and we watched as a huge container was loaded with animal slurry, chicken-flesh pink, pouring down funnels and pipes. We shot videos and photographs until the container was full, the waste bound for dog food.

A slaughterhouse worker shut the roller door. Mary-chris filmed until it fully closed. The lane stank, but not as noxiously as I'd imagined. In fact, neither the smell nor the confusion and fear in the eyes of the chickens disturbed me most. What shocked and depressed me was the mechanization of the process and the vast numbers involved. It was the first time I'd seen this up close. I wanted to cry and remain fearsomely dry-eyed.

The vigil was nearly over. Before Mary-chris headed off she gave me a hug and told me that my coming out in the rain and all the way from the United Kingdom had made her day. I'd landed only the night before and didn't know my way around by public transport, so I left the vigil as I arrived—in my running gear—and ran another eleven miles around beautiful English Bay. I needed to keep moving my body, because if I stopped I would begin to shake. I felt confused, shocked, sad, and, I understood, *vulnerable*.

—

Vulnerability is commonplace within activism of all stripes. Mark Hawthorne writes in *Striking at the Roots* of the "emotional pushback" of dealing with the "grief, depression, anger, stress or that psychic demon known as burnout" (226) that compassionate work engenders, and that turns people away from "coming to care." But this wasn't the vulnerability I felt. Mine was a sense of uncomfortable identification—of being *seen* for the first time on "the animal side" in such a direct encounter.

And this was not only to be seen by the commuters, but by the chickens: my frightened, confused, short-lived companions. I felt the power that Jacques Derrida assigns to the gaze of his cat: the look that makes us aware of "the naked truth of every gaze, when that truth *allows me to see and be seen* through the eyes of the other, in the *seeing* and not just *seen eyes* of the other" (12) [emphasis in original]. It is through the other that we identify ourselves. To act differently—to become part of a slaughterhouse vigil—would be an act of counter-identification. It would begin, as Anat Pick puts it, a reversal of "the untold labor that goes into the sustaining and upkeep of identity" (85), an identity that had so far in my life *not* placed my body on any front line—let alone one for such an "insignificant" creature as a chicken. Until now I'd kept myself safely hidden away: theoretical, intellectual, rational . . . *in*vulnerable. To identify *with* the chickens would be to ask hard questions of my own sense of significance, as well as theirs.

As I stood with the chickens, I began to understand why the image of Babe haunted me so much. To *not* take action is to deny our agency already at work in creating the conditions under which animals suffer. As Pick argues, "to deny the agency of creaturely suffering does not only deflect a difficult reality, but compromises thought" (11). The thought it compromises is the dominant epistemology of thought itself, a disentangled and sanitized rationality. Bodily inaction leaves us, perhaps, with only thought. As Bruce Friedrich, former director of policy at Farm Sanctuary, writes in the foreword to Hawthorne's *Striking at the Roots*: "It is not enough to withdraw our support for cruelty. If it were, the most ethical lifestyle would be that of the hermit in the mountains" (11). Or an academic in a tower. Rather, says Friedrich, we must go further: "Refusing to support cruelty and suffering is crucial, but the next *step*—resisting injustice—is even more important" (10) [my emphasis]. Friedrich's words are an example of Lakoff and Johnson's idea that our cognitive acts are modeled on metaphors of movement: the *step* into, and against, injustice we find in the world. Taking a *stand*. And so I stepped onto the line. Identity follows action.

Changing the Narrative on Animals

I'd almost wimped out of meeting Mary-chris that morning. Jetlagged and doubting my decision not to work with VanAqua, the inner script of myself as someone no good for frontline activism reasserted itself. Wasn't I the same frightened person who'd turned away from joining protests back in the United Kingdom? Couldn't I *just* be a writer? By the time I had my running shoes on, I was able to think more clearly (and running clearly would become a theme of the trip). What would I be giving up if I didn't go?

The step gave me a new story: a new script of who I could become. It was only the first step, and a small one, but I was inspired nonetheless. Stories are perhaps the most effective way to reach people and have them reconsider their personal behavior as well as public policy, and they've been used by large and small organizations for a long time, as well as by novelists such as Charles Dickens and Leo Tolstoy, among others. Over twenty years ago when he was chief knowledge officer at the World Bank, Steve Denning popularized the "springboard story"—an anecdote that makes a larger, more complex issue more comprehensible—as part of his work to communicate economic ideas. The TED talks have perfected the art of storytelling in twenty minutes to spread knowledge. In the nonprofit world, whole conferences are dedicated to developing "soap-opera" stories, while Nick Cooney's *Change of Heart* emphasizes the power of the single tale for activists working on campaigns to change behavior.

George Lakoff and Mark Johnson argue that the way we conceptualize our mind is through metaphors of the body: of manipulating objects, moving, eating. Our concepts for how we think and what we think with are bodily experiences when parsed through language: our minds *wander*, we *get the picture*, we have a *hunger* for knowledge, we *regurgitate* ideas. And so language, stories, narrative, and literature, made up of such metaphors, can be the catalysts for exposing the ways in which our reason is tied to the body, ours and others. The cognitive sciences, Lakoff and Johnson continue, "tell us that human reason is a form of animal reason, a reason inextricably tied to our bodies and the peculiarities of our brains" (17). Add such scientific research to our innate love of stories, and you've got the potential tools to change the world one listener or reader at a time.

The stories we tell about nonhuman animals have been repeated so often that we forget they are merely stories. Cows and sheep are generally (and wrongly) perceived in our culture to be passive, docile, often stupid, animals. These are not good traits for the protagonists of dramatic narratives. Who looks at the cow's "dumb" stare and understands in fact this is how she is subtly communicating to her companions?

The pig, on the contrary, has a long and powerful symbolic history, in both bucolic fantasies of farming life and in the disruptive, anti-speciesist narratives that have emerged since (and before) Upton Sinclair's *The Jungle*. And while many things have been said about the pig, they have rarely been labeled docile or stupid. The pig, or boar, or swine, has stood for chauvinism, stubbornness, and cruelty, as well as slovenliness, filth, and intemperance—all undeserved markers of character for clean, sociable, intelligent beings.

In the Bible, the Prodigal Son suffers the indignity of descending to the level of a swineherd after he loses his father's inheritance (Luke 15:15–17). Jesus allows the evil spirits that torment Legion to take up residence in a herd of swine, who then rush into the sea and drown (Luke 8:26–37). The pigs in these biblical narratives are innocent in both cases. They serve as reflections and repositories of the distress and humiliation of a human being. In this way, the pig not only represents aspects of animal life back to us, but defines, as historians Robert Malcolmson and Stephanos Mastoris put it, "a boundary between the civilized and uncivilized, the refined and the unrefined" (2).

Peter Stallybrass and Allon White have written in their study of the concept of transgression that before intensive animal agriculture put the pig into warehouses and gestation crates, she was kept on smallholdings, even in the home, fed on swill and scraps, and became the "creature of the threshold" that, they say, "overlapped with, and confusingly debased, human habitat and diet alike. Its mode of life was not different from, but alarmingly imbricated with, the forms of life which betokened civility" (47). As Babe stood on the road after her defiant leap, she was also standing at this boundary—touching both sides. She had leapt from the uncivilized into the civilized world to become a (some)body whose chance at life was recognized. In this gesture, she became the animal most like us.

The pig appears throughout literary history, such as in Shakespeare's *Richard III*, where Richard's adoption of a wild boar as his livery badge symbol is used by his enemy Richmond to describe Richard's aggression and destructiveness (act 5, scene 2, lines 5–15). Yet from Orwell's *Animal Farm* to E. B. White's *Charlotte's*

Web, not to forget *Babe* the film (based on the children's book by Dick King Smith), pig texts are far from always being metaphors for the difficulties of human agency, and readers are coming to accept that books and films that feature animal protagonists are *about* animal bodies and their experiences, and not only allegories of human relations.

As Raymond Williams identified back in 1971, Orwell wrote *Animal Farm* after witnessing the violent beating of a horse by a farm hand, and began to reimagine that abuse through a Marxist lens and not, as is commonly thought, reimagining the Soviet Union through the metaphor of farmed animals. In the 2014 novel *The Awareness*, Gene Stone and coauthor Jon Doyle chose four protagonists to tell the story of the world's nonhuman population coming to conscious awareness of their exploited positions. The four protagonists are a wild bear, a domesticated dog, a circus elephant, and, for the farmed animal population, pig number 323. Robin Lamont's new Kinship series begins with *The Chain*, a narrative that investigates corruption in a pig processing facility. The pig's historical closeness to us, her established intelligence and perhaps even the pinkness of her skin, make her a useful protagonist in stories that question and trouble our relationship with animals. The cultural theorist Susan McHugh adds: "The terrific significance of pig parts in modern fiction . . . indicates a continuing fascination with stories of this animal as an exceptionally versatile symbol, if not messmate" (170).

Literature is a form of writing that at its best draws attention to how we construct meaning through language and in so doing reveals the fact that our metaphors for reason and rationality, on which speciesism is based, are merely repeated stories. An excellent example would be Margaret Atwood's reimagining of the post-farm future of pigs in her MaddAddam trilogy, beginning in 2003 with *Oryx and Crake*. In her dystopia, pigs have been genetically modified and, with even greater intellect than they possess now (pigs are thought to be the fourth most intelligent species after humans, primates, and cetaceans), they learn to live ambivalently side by side with the remaining human population, often targeting *us* for their next meal.

In Marie Darrieussecq's novel *Pig Tales* (*Truismes* in the original French) the unnamed narrator, a woman whose body turns into that of a sow, experiences what it is to *become* a pig. Her perceptual experience of the world and herself changes. This conceit has an ancient provenance. Homer recounts the story of the witch Circe, who turns Odysseus's crew into swine. In Plutarch's *Gryllus,* a pastiche of the Circe episode written in the first century C.E., Odysseus needs his crew back if he is going to return home to Ithaca and tries to persuade them to return to human form. Gryllus, a pig, thinks otherwise: "Now you come and try to persuade us, who have all the good things we want, to give them all up . . . and to sail away with you, reverting again to the condition of man, the most miserable thing alive. . . . Animals have souls that are more naturally and perfectly disposed to produce virtue than men have."

In *Pig Tales*, the more the narrator becomes a sow, the greater is her attention to her place in nature: "The air, the birds—I don't know, whatever nature was left—really affected me all of a sudden" (10). Later on, when more fully in pig form and wild in the forest, she foregoes all human concerns:

> All the winter of the Earth exploded in my mouth. I no longer remembered either the millennium to come or any of my experiences—everything rolled up into a ball inside me and I forgot it all. I lost my memory, I have no idea for how long. I ate and ate. . . . Eating and sleeping didn't require much effort, just a little vital force, and there was enough of that in my sow's brain and vulva and brawn to wallow through life. . . . With my entire body I felt once again the spinning of the planet. I breathed with the shifting winds, my heart beat with the surging tides, and my blood flowed like a torrent of melting snows. I flexed my muscles in communion with the trees, odours, mosses, ferns, and rotting leaves. I felt the rallying cry of the animal kingdom course through my body—the ritual combats of the mating season, the musky aroma of my race in a rut. A craving for life sent shivers through me, engulfed me. (127)

When the narrator becomes upset by environmental factors, such as the threat of "leaving my lair," then it is her body that responds: she becomes more pig, or rather, "I was naturally rather upset the day we moved, so I was a *complete* pig" (121) [my emphasis]. The narrator experiences the full extent of her affective life (as we do when threatened) in her animal body. That is, both she and we attain more complete experiences of our lives when we feel ourselves as embodied creatures.

But the transition to animal is not as easy as becoming more fully embodied. The things of culture are designed to lead us away from our bodies, which is what happens to the narrator. These things of culture in *Pig Tales* are most notably acts of reading and writing. Anat Pick observes: "*Pig Tales* begins by making the relationship between species and writing one of its chief concerns" (84).

From the beginning, the narrator consciously draws attention to the process of recording her story in what the English translation terms "piggle-squiggles" (1) or *écriture de cochon* in the original—the writing by/for pigs. The narrator is also writing as a pig: with difficulty, for example, holding the pen in her trotter. The term *écriture de cochon* resonates with *écriture féminine,* the term coined by Hélène Cixous in *The Laugh of the Medusa,* meaning "feminine writing" but that also opens to the inscription of the female body and female difference—indeed, which, like Circe, echoes with the monstrous, mythic *femme fatale et animale*—of language and text.

By toying with this idea, Darrieussecq has written a book, says Pick, "about a certain animality *of* writing that exceeds the stakes of the female writer" (81). That is, the idea of an *écriture de cochon* of writing by and for a pig reveals the naturalized ways in which writing/language normally supports and perpetuates dominant (anthropocentric and androcentric) norms. Such "piggle-squiggles" from a sow/woman expose the ways in which both female and animal bodies are inscribed with difference through language and text, at their subordination and expense. Focusing on the way language works shows, Pick continues, that "Darrieussecq is interested in how identities (human, pig, woman, whore) linger, travel, and seep into one another, flickering, as it were, in their own difference, their own becoming" (85). To have another species—the pig, a sow—be a writer challenges the idea, at least in the text, that only humans write. If only humans write, then language can be part of what it means to be human and is implicated in the process of becoming human. For Pick, "the process of 'becoming-human' [is] for Darrieussecq inseparable from the question of writing" (85). To ask the question of the relationship between species and writing is also to expose and explore the "writing of species." It is to suggest that the differences between species are *merely* written; are merely, then, a question of language and not something *natural* at all.

It is the seemingly human act of remembering her lost partner that begins again to separate the narrator from her embodied experience of pig life. An emphasis placed on human thought brings her back to her mind as separate from the body: "It was the thought of Yvan that roused me. Pain flooded back into my belly, and I came round. Frightened of losing myself completely, the way I lost Yvan, I tried hard to stand upright. That hurt" (127).

The narrator tries to be bipedal: to be human. She cannot quite let go of her human identity. She is frightened of the experience of living a fully embodied life and losing the victories of the human ego in carving out for her (as for each of us) an identity that fits within the acceptable forms of human culture—forms that have for the last few hundred years, if not thousand, expressed the primacy of rationality over embodied emotionality, of man over woman, of human over animal.

It is hard to give up such gains ("That hurt") that bring us what our culture says we most want. The civilizing process produces seemingly stable, autonomous, bound, and single identities. But civilizing is also traumatizing; we are no such things as stable, autonomous, and bound individuals. Seeing and believing ourselves to be so, we suffer all sorts of atomization and pathology as we move away from the things we actually most need: touch, somatic and physical regulation, organicity and symbiosis with our environments. What we require as a species to flourish is to experience life with our vulnerable bodies—to accept our material corporeality. We must do so without what Anat Pick calls "comfort thinking," which she suggests is "perceiving the body with consolatory illusions" (186). These comforting illusions of the mind ("thinking") are that we operate as autonomous, separate, nonmaterial, and invulnerable beings, and cannot be threatened by the material, animal world.

In David Mitchell's novel *Cloud Atlas*, the clone Somni-451 comes to fully face her own consolatory illusions when she learns the truth that her species, her breed, are nothing but slaves. When their useful service lives in fast food restaurants (serving meat burgers!) are over, rather than receive the paradisiacal promise of "Exultation," the fabricants are liquefied into food forms and fed to the newer clones. Through a metaphorical (and, in the film version, a literal) curtain, Somni-451 sees fellow fabricants suspended from chains; the scene is modeled exactly on the contemporary slaughterhouse.

The coming to awareness of Somni-451 could be considered the "coming to care" that Elspeth Probyn suggests we need to rebalance our relationship with the nonhuman. It is the same for Pig 323 in *The Awareness*, as the world's animals realize their enslavement. This revelation is the mirror of ours when we come to care and understand the role we have played in the treatment of nonhuman animals. It also suggests the thought experiment of an alien race that treats us as we do nonhuman animals (a proposition that Sue Donaldson and Will Kymlicka put forward in their book *Zoopolis: A Political Theory of Animal Rights*). Such is the premise of Michel Faber's novel *Under the Skin.* A sophisticated, intelligent, technologically advanced alien race arrives on Earth and kidnaps a few unlucky human victims and consigns them to underground slaughterhouses as meat animals. The victims ask for mercy with language and gestures that the alien race understand but choose to ignore for their own benefit. Faber invites uncomfortable questions of our own willfulness in ignoring the cries for aid and mercy from those we, as a species, currently exploit.

Ultimately, these novels ask how our identities alter—radically—as human beings when others perceive our bodies differently. This is the position that Val Plumwood famously found herself in when taken in the jaws of a crocodile: she became *meat.* Plumwood used this experience to trouble species boundaries and to reconfigure our understanding of selfhood as human. Such a realignment might not automatically happen with the pigs and cows and sheep and fish whom we abuse and exploit, since they are not predators. But we would, Plumwood argues, find our corporeality much more real to us and our material relations with others would change.

The power of cultural systems is made clear in *Pig Tales.* One of these systems is the process of slaughter. It is while she is packed in the back of a truck on the way to an abattoir that the protagonist returns to human form. That reverse metamorphosis, as it were, is what saves her, because with her reconstituted humanity she has access to her human mind and perceptual system and is able to recognize and unlock the door of the truck to escape (perhaps the author did not know about the eighteenth-century pigs unlocking their gates). This scene emphasizes the (ir)rationality of the slaughter of animal bodies. Such an act can only be rationalized from the site of the disconnected human and *not* from the bodily entanglement with the pig. The novel also reinscribes the notion that to escape is only a human activity. Babe tells us otherwise.

Slaughter also emphasizes for us exactly why we desire to be separate from nonhuman animals. The pig on her way to slaughter is almost universally impotent and vulnerable—all save the very few, such as Babe, who express their desire to live. We have done all we can to distance ourselves from this type of absolute vulnerability. Modern killing, however, is an *in*escapable part of our species identity. As Jonathan Burt writes in *Killing Animals*, industrial slaughter practice operates as "one of the constituting elements of our particular social identity" (122): the mechanization, "efficiency," and mass disassembly lines that compartmentalize animals, workers, emotions, and bodily parts. At heart, the modern slaughterhouse encapsulates the idea that, no matter how powerless we may be, we are always more powerful than the eviscerated. Speciesism rests on this (mis)conception of ourselves: that we are invulnerable, that we are not material, and that our bodies are not subject to more powerful external forces. We project our fears onto the bodies of others—mainly other species, although far from exclusively. We do so because we are afraid of our own bodily mortality.

Direct Action and Narrative Strategy

Before leaving Vancouver, I met with another activist group. Direct Action Everywhere (DxE) is a relatively new San Francisco Bay Area–founded grassroots animal liberation network. Inspired by the civil rights and anti-slavery movements, DxE's mission is to use "creative nonviolent protest . . . to tell the animals' story." The members' model is one they hope to spread easily—anyone can download their materials to stage a protest as part of a worldwide community. Their speaking and protest tours around North America are part of this mission and I went to join them for an evening in the second-floor offices of a social justice collective above the Writers' Exchange on East Hastings Street, only a block from the Liberation BC chicken vigil.

The main target of DxE actions is the fast food chain Chipotle, which has become the new poster child for success in corporate America, with a market capitalization of over $17 billion. My research into what a narrative organization might be—one that models storytelling from top to bottom as a means to make change in the world—appears to be exemplified by Chipotle. Their marketing vision is a "narrative of change" about how people consume fast food. Chipotle knows what it's doing—the restaurant chain's short film *The Scarecrow*, where a small-scale traditional farmer takes on "Big Ag" and the use of GMOs, won top prize in the PR category at the Cannes Film Festival, painting Chipotle as the good guy and asking consumers to join them in "the quest for wholesome, sustainable food." Chipotle has spent millions paying storytellers to write marketing and packaging materials. One of these narratives claims that a farmer from whom the chain gets its pig meat communicates with his animals telepathically—that's how he knows they're happy. I wonder if telepathy passes through the thick walls of a slaughterhouse, or if the farmer's feedlot is out of range when one of the flies in the ointment of the "humane meat" story—the slaughter—lands and gets stuck.

Wayne Hsiung, cofounder of DxE, calls Chipotle's marketing a fraud. "Chipotle has the audacity to claim that it is killing with 'love' and 'integrity,'" he told me. "It invested millions into *The Scarecrow* with idyllic scenes of humane and sustainable farming—but no images of slaughter. More than any other restaurant, Chipotle has tied its fortunes to the myth of 'humane meat.'"

Much of DxE's activism focuses on exposing what they see as "lies" sold by Chipotle as marketing. So DxE invests in its own narratives that un-truth some of those "truths." The presentation that evening in Vancouver was all about "meme spreading" and "visual storytelling." This form of direct action against Chipotle was what

I'd come to find: a battle of narrative and counter-narrative played out on the field of animal rights. And so, heartened by my first contribution with Mary-chris at the chicken slaughterhouse, I agreed to join DxE on an action ten days later, on the last day of the annual Animal Rights Conference, held in 2014 in Los Angeles.

—

It's 9 p.m. and I'm lying with thirty others on the floor of a Chipotle restaurant off El Segundo Boulevard in Los Angeles. I open one eye and squint. Wayne and other core organizers are filming the "die-in," which began when a hoody-wearing activist shouted the simple tagline "It's not food! It's violence!" The thirty of us sitting or standing around the restaurant all dropped to the floor. The manager, a young woman in black, skitters around us, talking in a low voice to Wayne, asking him to get us to leave. I close my eye and listen to the music.

We're on the ground for about five minutes, maybe less. I hear the shouts of a customer. He's calling us dicks. He says what we're doing "only encourages people to eat more animals." I'm trying to figure out why making visible the violence in animal agriculture would encourage people to subscribe to it more deeply. I think of the theoretical work around threatened identities, and here it is: the reality of what happens when you put pressure on someone's sense of continuity, their belonging to a social structure that privileges their taste preferences over the bodies of nonhumans.

Then we're up. The cue is the resumption of the chant "It's not food, it's violence!" and we all file outside. We face the restaurant holding up signs. I feel sorry for the guy sitting outside on his own, eating his sandwich and fries. I stand somewhere near the back, a little shy. Other members of the crowd lead chants about the desire for animal liberation and the injustice of low-paid service jobs. The manager is at the door on her walkie-talkie, asking Wayne to come back inside.

Something happens. A Sikh man comes and joins the protest. He stands at the front in his clean-pressed white shirt and gray trousers and holds up one end of a placard carried by a protestor. After a few minutes, he goes and gets his family and leaves without buying anything.

Then something ugly happens. The guy who'd been heckling us runs out. He's about 6 feet 3 inches tall, in a black T-shirt and jeans, with a wispy blond ponytail. He's high on anger. "Kill more animals! Kill more animals!" he shouts, pumping the air. I see emotions swirling around him: indignation, fear, but also enjoyment in being riled. Unkindly, I think this is the most exciting thing that's happened to him for a while (*and isn't it for me?*). Who the hell wants to "kill more animals!"? For a moment, I imagine Chipotle and its customers as the Philistine Goliath to DxE's young David: "And I will strike you down; and I will give the dead bodies of the host of the

Philistines this day to the birds of the air and to the wild beasts of the earth," says David (1 Samuel 17:46). I see the chickens and cows and pigs joyous in the slaying of this modern-day giant.

The police arrive. Our on-the-spot attorney goes over and preempts their actions while we file away to our meeting point across the parking lot. The protest is over and we disperse; we make our way to Veggie Grill for some food.

That wasn't so bad, I think. We may have changed one person's behavior; entrenched that of another. For the rest, who knows? To "think through the body" about such a form of protest would be to consider it an *enaction* of exposure—exposing those of us taking part and also the ways in which the system works. It was exposing for those witnessing and for those who felt threatened—making us vulnerable, a visceral and inescapable opening up of the stories we tell both about nonhumans and about our identities. It seemed to me that DxE's strategy was to help people understand they live in a world made up of stories, about themselves and the food they eat; the simple narration of another perspective ("It's not food, it's violence"), when combined with the confrontational aspect of the event, could be a catalyst for change. These protests are forms of bodily movement and bodily risk—exposing us to vulnerability.

The problem was this: I'd seen that kind of response before from the man who wanted to "kill more animals."

It was the response I saw to the high street protests back in the United Kingdom: of people backing away from activists shouting at them through megaphones; or in actively confronting the animal rights message with the opposite. It was why I'd turned away from activism; why I almost didn't join the "die-in" but, as the cars left to head to El Segundo Boulevard and the restaurant, walked away—the act of a frightened but also frustrated person. The protest felt, even behind a mask, *too* exposing. I'd imagined it wouldn't do any good nor bring people compassionately and effectively over to the animal side. Perhaps its uses are elsewhere—in the materials that are circulated online or in the energy and passion such work validates for activists. These protests are perhaps most of all for the actors. One DxE supporter, Sarah, told me that she worked in business consultancy but had only really "found her voice" speaking out for the animals. Although nearly as new to protest as I was, she'd been at the front of the group leading the chants.

I was proud of myself. Even so, the reservations I'd felt at taking part were resurfacing. This was, I realized, because I now had something else to compare such protests to. It seemed that the welcoming smile of Mary-chris, offered as an invitation to passersby outside the chicken packing plant, was more powerful—and would be in the long term more effective—than confrontational forms of protest. More importantly, the action with DxE lacked the encounter with the living nonhuman beings who were subjects of their own lives; whose eyes I could look into and be "seen seen." It was this welcoming invitation into a tragic direct encounter that, from my perspective at least, offered greater possibility of reimagining and—critically—re-embodying our relations to the nonhuman other. I wanted to experience more of this kind of bodily act.

—

Losing sight of the bodies of those animals with whom we are in Blackman's "brain–body–world entanglements" is a willful blindness that Derrida accuses us of consenting to in our refusal to be "seen seen" by the animal. This refusal is to be complicit with the structures that distance us from the animal—including the structures of language that we use to "write other species." Our stories must re/spect—look again, more closely at—the nonhuman. When we do look again, as have the authors addressed above, we can see that our stories of experience are always entangled with those of other species. Such stories are embodied in our turning toward the animal, too, through direct action, bearing witness, and the act of building the corpus of animal rights stories across the world. So to answer my own question: could I not *just* be a writer? The answer, for myself at least, was *No*. I needed to move and mobilize my body in different ways, across new terrains, and in alliance with other bodies if I were to discover the stories I need to tell.

Stories, however, carry risk. We are described in but *not* constituted by language. And we've all heard the cry that "we need new stories" too many times. Sharing new narratives does not straightforwardly rewrite who we are. Rather, the stories can help expose the means by which old narratives work to reinforce norms (through repetition). Stories are only useful to us if such exposure halts "deflecting" the real experience of our bodies onto and through language. Otherwise, the story becomes another tall tale, a metaphor, an allegory. If we *do* learn new scripts in our bodies, we can then see ourselves as animals and think of ourselves as animals. Become (again) animals. We must always bear witness to the body. As pattrice jones, cofounder of VINE sanctuary and author of *The Oxen at the Intersection*, reminds us, we must always ask: "*Where's the body?*" (12) [emphasis in original].

Locating the body of and within the story is far from easy. As in all struggles, animal liberation will succeed when it finds a story we can come to accept as "normal" and "right." Through stories we can make friends with the knowledge that we are vulnerable. Such knowledge, if we know it at all, must stem from corporeal materiality. To do this entails becoming aware of the fact that the pig, cow, duck, or sheep is an embodied creature, too; and that their minds are also embodied. They know, perhaps much better than we, that cognition is a bodily process.

5

The Sanctuary as Embodied Place

WE CAN LEARN to care for nonhumans in many places. Animal sanctuaries are one of the exemplary spaces where, especially for those new to engaging with nonhuman others, questions of veganism and animal advocacy can be asked without the overly discomfiting exposure of "being seen," of feeling as if one is challenging the dominant meat-eating culture. It's a chance to come closer to the bodies of (previously) farmed animals, to experience their personalities and be their guests. Farmed animal sanctuaries are where, says Pam Ahern, founder of Edgar's Mission in Australia, writing in *Turning Points in Compassion*, people can have their "feelings and concerns for animals validated" (68). (Edgar's Mission is the sanctuary where the pig stars of both *Babe* and *Charlotte's Web* lived out their days.)

Sanctuaries are often microcosms of vegan worlds where all species are considered equal. (In fact, not all species, but most: worms and parasites that plague larger mammals and groundhogs and rats that dig into the food supplies are some examples of unwelcome visitors, and continue to trouble the lines across which species are kept out of even these gentle places.) The sanctuary remains a place organized by humans, necessarily so because of the broken bodies, infections, trauma, and disease affecting many of the animals who arrive at the sanctuary.

I noted that sanctuaries are safe havens for would-be activists who don't want to be exposed to the critical attention of other human beings. They will, however, be seen. Just as the chickens *saw* me at the slaughterhouse in Vancouver, visitors to farmed animal sanctuaries will be seen by the resident nonhumans. In the experience of many, the identification is kind and forgiving; although it was not, I found, wholly free from discomfort. Even in such places, the "consolatory illusions" of comfort thinking are challenged.

During my visits to sanctuaries I've often thought of Babe and her fate as the mascot of the Foshan police force, living out her life as a solitary pig on the concrete floor of the police dog enclosure until, at least, the police grow bored of her. For Babe to reach a sanctuary would have meant proper treatment, the company of her kin and other nonhuman animals, and the ability to express her natural desires to root, forage, bathe, and sleep.

As part of my fellowship I wanted to visit the animal rescue and education organization Farm Sanctuary. After reading cofounder Gene Baur's book *Farm Sanctuary: Changing Hearts and Minds About Animals and Food*, I wrote and asked him whether I could visit to see how sanctuaries worked to change identities by bringing the bodies of humans and nonhumans closer, in non-exploitative settings. I thought I might bring the writer's skill set of questioning, narrating, and imagining to Farm Sanctuary's aims and mission. I hoped to see Babe's brothers and sisters expressing themselves and to know that somewhere on the planet at least a handful of pigs were living free from pain and fear. I wanted to find a way to fight back in the war against animals—and to do so in a place where it is easier to console oneself that the fight is not impossible and that, just perhaps, it might one day even be won. Not all comfort thinking is bad for us.

Living a Sanctuary Life

On a warm, sunny afternoon at the end of July, after a six-hour drive from the Toronto airport, I pulled into Aikens Road, a dusty trail about fifteen minutes outside Watkins Glen at the bottom tip of the Finger Lakes in New York State. A little way along, I stopped the car and got out to get rid of my belt.

It was leather. I'd bought it six years before. No meat or dairy products are allowed at Farm Sanctuary and no wool or leather is to be worn. I'd been meaning to replace the belt for a long while, as I already had with my shoes and wallet. I'd poked around the thrift stores and found nothing suitable. Belt off, my jeans were hanging down. I was in the middle of pristine country. I couldn't just throw it away, could I?

These were excuses, of course. Holding on to the belt was much more convenient than doing the work to get a new one. It bound me to a social identity that I, like the narrator in *Pig Tales*, was not able to fully abandon (just yet)—one I'd belonged to for much of my life. Disposing of the belt would, of course, move me further toward the new identity I was seeking. But questions of identity mixed with a little vanity are a strong, strange combination. I didn't want to look scruffy. I knew I should have done this before, I berated myself, knowing I was angry at myself for much more deeply troubling behaviors.

I stored the belt in the trunk of the car and drove on. Five minutes later, turning into the drive, I forgot all about it as the feeling that this was where I was meant to be flooded over me. It was as if I'd come home.

"Well, good. It's a sanctuary for people as well as animals," Gene Baur said as I told him about my feelings on driving up. We were taking a long walk around the grounds. "It's a very peaceful place," he said. "Being more compassionate about the way we treat others is what this place is about. People think that's only about the animals. But of course it becomes about how we treat each other, too."

Farm Sanctuary currently looks after around seven hundred resident animals across three shelters and re-homes thousands more each year. It has a permanent staff and a phalanx of volunteers, donors, and supporters. I understood the feelings Farm Sanctuary and its nonhuman residents stirred in people. The atmosphere was of a peaceable kingdom (Isaiah 2:2–4) with its vision of swords turned into ploughshares. An hour touring the sanctuary with Gene was a gift. He was modeling what was required from visitors: consideration and patience. I felt a strong sense of welcome. Farm Sanctuary, he was telling me, is a trusting organization, made up of trusting beings—nonhuman as well as the staff and visitors.

As we walked and talked, I asked Gene what it was he brought to Farm Sanctuary, and why he felt it had been so successful as an organization.

"Vision," Gene told me. He was the big-picture guy, the "dreamer." When he and his former wife and cofounder originally bought the Watkins Glen acreage in 1986, Gene was able to look past the ramshackle state of the place and envisage what it might become. Gene knew such a vision would be tough to implement. But the work of animal activists is to reform the relationship between species, which takes hard work. I'd come to explore the importance of sanctuaries for the advocacy movement, and in particular to learn about the ways in which personal contact between species works to change that relationship. Most farmers, after all, know the bodies of their animals, but few exhibit Lori Gruen's "entangled empathy" to consider them worthy of their self-directed lives as equals. What I wanted to learn was how sanctuaries exemplified the practice of changing exploitative relationships, and to do so based on bodies becoming trusted guides on a path toward flourishing.

Susie's Story

For the past fifteen years, Susie Coston has been the national shelter director of Farm Sanctuary. It is her job to oversee the care of all the nonhuman animals—residents and those that pass through on their way to homes. Since the great majority of animals who arrive at farmed animal sanctuaries come from industrialized farms or stockyards, they suffer from complex injuries, serious illnesses, and physical deformities, which Susie and her team deal with daily.

When I met Susie on her late shift she was suitably unkempt for someone who'd been at work since 8 A.M. because there was so much to do. Susie had hand-raised many of the animals on the farm. "Even now I still do the caregiving," she explained, "at least once or twice a week. These animals know me, we have a bond." This bond was maintained by, or rather in constant "becoming-with," their proximity.

Out on the farm our first stop was the chickens. A few years ago, Farm Sanctuary decided it would be a better experience if the chickens were allowed to run free rather than be kept in a shed. So they lifted the fences and gave room for the chickens to come in and out, and the experience for both visitors and chickens was transformed. The chickens are extremely friendly, explained Susie, and come running when anyone appears. They're also able to exhibit all of their chicken-specific behaviors in the open, such as dust bathing. The chickens recognized Susie; when they ran, they ran to her.

"It's a total experience getting a tour from Susie," Krysta Vollbrecht, who works at communications for Farm Sanctuary, told me. "With Susie, you see that the animals know her voice. Know *her*. And trust her. When Susie's there you get to see the full range of their behaviors. You get to see them being who they are."

Susie and I walked into the feed shed. "Here," she said, pouring seeds into my cupped hands. "Miranda doesn't like to eat off the floor because she's been de-beaked, and it hurts her. She prefers to eat out of something soft, like your hands."

I bent down to let the shy, somewhat withdrawn Miranda take her meal. The pecking movement was forceful, although it didn't hurt my palms. I could see why eating off a hard floor would be painful.

"Chickens and turkeys, they don't have fingers, right," said Susie, "so they explore the world with their beaks." Indeed, the sensitive beak contains more nerve endings than a human fingertip. "They're curious, too. To cut off a bird's beak just makes no sense." Susie told me that industrial agriculture de-beaks the birds, always without anesthetic, to stop them maiming and killing each other when they fight, which they do because of the terribly cramped conditions in which they are kept.

In a nearby pen was the newest arrival—a young cow, Chandini. Before she arrived, the rescuers told Susie she was a calf, but Susie inspected her teeth and discovered she was at least a year old.

"She's tiny," said Susie, as we watched Chandini eat. "You shouldn't be able to see her jaw. I'm worried, because if she's this small, then there might be something wrong with her."

Chandini had come from New York City, where a Hindu family had bought her as a gift for their priest as part of a religious celebration. However, the priest lived in Brooklyn and couldn't take a cow. The family panicked. They rang around, and finally Farm Sanctuary agreed to take her in. I asked Susie what she thought of the family who purchased Chandini.

"Makes no sense, right?" she answered.

We hiked to the top pasture to move the other cows, who'd been mooing since the previous evening, when they ran out of grass to eat. As we approached, the angle of the slope hid the size of these cows, but now I could grasp just how enormous they were. The back legs of the black and white Holstein Friesians were perhaps eight foot in height. I'm unsure why I didn't know that cows grew this large, and commented on this to Susie.

"That's because they're dairy," she responded. In the dairy industry, male cows never grow to maturity. "Beef" cows, which can still weigh in at over a thousand pounds, are smaller and have much shorter legs. But these Holstein Friesians were as big as oxen. Seeing them brought home to me how impoverished our contemporary experience with the actual bodies of living farmed animals is—an encounter that would've been common before the Industrial Revolution, when farming employed most people. Nowadays, we don't know who these animals are or what astonishing creatures they become when they're allowed to survive into maturity and old age.

The tallest was Jay, easily recognizable because of his fire-scarred back. Jay was in a transport truck carrying thirty-four cattle when it crashed into another vehicle and burst into flames. Eighteen cattle died in the wrecked trailer. Others collapsed on the road and lay dying from their wounds. A second truck soon arrived to take the survivors to their original destination—the slaughterhouse. All of those still on their feet were rounded up, except one two-year-old Holstein steer who was horribly burned but determined to live. Like Babe, he took off. A twelve-hour chase ensued before Jay was finally captured and taken to an animal shelter. Local residents campaigned for his life to be spared and custody was relinquished to Farm Sanctuary.

As we looked up at the cows waiting by the gate, Susie shouted to one of them: "Sonny!" Immediately, the cow replied with a bellow. Susie laughed. "I bottle-fed him," she said. Sonny's story is depicted in Jo-Anne

McArthur's book *We Animals* and Liz Marshall's film *The Ghosts in Our Machine*—perhaps because of his gregarious personality, but also because what happened to him is emblematic of the callousness of industrial farming. Sonny was less than a day old when activists from the sanctuary found him lying in a filthy stockyard pen. As a male calf, he was useless to the dairy industry, and most calves are sold off at these stockyards for a few dollars as veal or cheap beef. His umbilical cord had been torn off, leaving an open wound. He was dehydrated, exhausted, and infected. Without intervention he would have died after a few hours of misery.

"Makes no sense, right?" said Susie.

This was the phrase Susie kept coming back to: that so much "makes no sense" when it comes to how nonhumans are treated. As full of laughter and smiles and compassion as Susie was, she made me wonder if this verbal tic was something of a safety valve in the face of so much suffering: that it makes no sense to push a chicken's body until it breaks (unless you see the chicken as a commodity); that it makes no sense to buy a cow for someone who lives in a Brooklyn apartment; that it makes no sense to leave a day-old calf to die alone and in pain.

How Sonny was treated reflects not the extremes of industrial farming but standard practice—animals as units subject to profit and loss. The sanctuary reverses that identification. It is Sonny's status as a being equal to everyone else that is at the heart of a sanctuary's basic work. And at the core of this work is Susie. Susie, it seemed to me, was engaged in the work of *sense-making*. Her job was to make sense of and out of human cruelty, not least by returning the senses to the traumatized nonhuman bodies she cared for. Such sense or sense-making lies at the heart of the relationship between bodies. It's a relationship that Susie has with these animals, and they with her. Making sense is a physical act, enacted as Susie wrestles with a goat to give him his antibiotics, or holds and soothes a traumatized chicken, or bottle-feeds an orphaned calf.

In economic terms, these relationships make no sense. In terms of restoring the senses and rejecting the numbing cruelties of factory farming, this work is eminently *sensible*. The experiences and knowledge of working with and caring for animals is a corporeal and material practice—one that, based on a recalibrated relationship with animals, produces a different kind of corporeal and material knowledge. For things to make sense, we must actually *sense* them. It is why Susie kept coming back to that phrase. For her, animal exploitation does not "make" sense. The opposite of sense-making is *trauma*. So sense-making can also be transformative—indeed, as an example of transformative justice, a relational and educational opportunity for victims, offenders, and all other members of the affected community to come to peace following trauma and violence.

As Lisa Blackman points out in *The Body: Key Concepts*, we are taught through our cultural practices and the broader dualism of the mind–body split to think of "sense making as a cognitive activity, rather than as a thinking, bodily, felt sense" (24). Yet relationship and entanglement only begin to *make sense* when we understand ourselves as sense-making creatures and recognize other beings as sense-making creatures, too. To be fully sense-making, we need to be alive to the rhythms and flow of those others with whom we are in contact. This contact is what sanctuary life provides. What I saw in Susie was a person living a more fully enriched, embodied, and sensuous life because of this opportunity to be cognizant of, or rather sensitive to, all her relationships.

In contrast, I saw the poverty of my own embodied experiences up to that point, having never spent much time with nonhumans in anything other than domestic settings or on the plate. This is perhaps why, when we do turn vegan, our compassion and understanding suddenly expand and we become defenders of many other nonhumans we'd been ambivalent toward before, such as the spider in the bath or the worm in the rain. Of course, even this compassion can butt up against limits, as when for some of us we encounter critters such as cockroaches and mosquitos, whom we see as our bodily antagonists. Is this where prolonged encounters with living others is no longer able to alter our behavior?

Bob Comis and the Last Pig

I still hadn't seen any pigs, so Susie took me to meet the current family. It was the pigs' sleeping time and most of them were buried under large piles of hay. Susie told me that many of the pigs from industrial farming become obsessed with the luxury of hay, having never experienced it—or mud baths or belly rubs—before. We entered their shed. The pigs smelled of poop and urine but the sty was clean, and more than anything the pigs' odor made me think of old, stale wine. The pigs and hay blended together in shades of yellow and white. But where the pigs were lying was unmistakable due to their great, barreling bellies and stocky legs that poked out of the bedding. When touched, their trotters shivered the hay like a blast of wind. We crouched down by the friendlier, less traumatized pigs and rubbed their backs and chins. I was surprised by how bristly their hair was and I thought of the image of Babe that had brought me here, across the world. I thought of all that picture had done for me in my journey, but also what I *couldn't* know from that image without coming closer to the actual body. Here, at least and at last, I was able to see, feel and smell, and experience what it was that Babe wanted from me. What I'd been called to her aid to provide.

As we headed back to Susie's office we met a few members of the communications team packing up the car to go and film a pig farmer who was "donating" five of his last pigs (all sick or injured) to Farm Sanctuary before turning his land into an organic vegetable farm. This was Bob Comis, and his was a story that also educated me on the terrors faced by the "Last Pig." His story helped make sense of what an act of courage it was for Babe to jump out of that truck at all. I'll come to that in a moment.

Stories of animal farmers having a moral awakening offer a valuable testimony for animal advocacy work. *Turning Points in Compassion*, an anthology of over sixty activist stories, begins with the section "Farmers as Visionaries" and lays out the transitions made by Harold Brown, Howard Lyman, and David Lay. They are also the backbone of the 2004 and 2009 *Peaceable Kingdom* films, and their stories have all of the fervor and satisfying trajectory of a religious conversion.

Bob Comis was a new addition to this list. He'd announced he was closing his pig farm because he'd decided that what he was doing was unethical. The next day, I was in the office when the communications team was looking through the unedited footage they'd shot of Gene and Bob. Gene and Bob discussed Bob's imminent change to a vegan farmer, something that had been on Bob's mind since April 2011, when he'd recorded his first concerns on his blog and was soon referring to himself as a "slaveholder and a murderer."

Bob told Gene that his awareness of the lives of the pigs had influenced his whole outlook and that he was now more compassionate and noticed things more. I was struck by how fit and strong Bob seemed: the result of managing a large farm and every week wrangling between two and ten pigs onto a trailer to take to his chosen slaughterhouses, certified "humane." His physical stature was not an insignificant element in the story: images of successful plant-based athletes and attractive, healthy individuals reduce the fears that some have that vegans or animal advocates are frail or unable to withstand the rigors of manual labor.

I noticed how patient Gene was in allowing Bob to feel his way through the maze of his feelings about the animals he raised for slaughter. Gene's approach to animal rights is staunchly to "meet people where they're at." His body exemplified this. He stood side-on and leaned back with hands in his pockets, all the while maintaining eye contact and a wide smile. He held this relaxed, unthreatening position to let Bob loosen up in the communicative space. Gene was not only present with Bob as he moved through his story, but he made sure Bob was as detailed as possible in his descriptions of his change.

The "last pig" was one of these details—the final animal in the pen in the "humane" slaughterhouse, after all of its family members and conspecifics had been killed. This "last pig," Comis observes, gives the lie to the idea of slaughter ever being "humane." As he has written: "Pigs live in groups not only because they find safety and comfort in numbers, but because they are intensely, and

I believe quite consciously, gregarious. . . . When those bonds are broken, a pig suffers a tremendous amount of psychological stress, and when the circumstances are right, pigs express that psychological suffering (stress is an inadequate term) of broken bonds by totally and completely flipping out. . . . You know where circumstances are just right for pigs to go dangerously ballistic? In pens at slaughterhouses. One by one as the day at the slaughterhouse passes, pigs are pulled out of pens with groups of pigs in them, until there is one last pig left in the pen."

Did Babe "flip out"? Did she, to use Comis's words, go ballistic because she feared that she would become the last pig? Was her leap not only a defiant savoring of freedom but the act of an animal who, frightened and alone, knew it was now or never? And what about the other pig in that picture? Did she feel the same way?

One other aspect in the video struck me: Gene and Bob knew they were being filmed and both presented themselves accordingly. Gene was affable, calm, and professional. Bob was much more animated and articulate about his feelings. But in his attempts to explain his journey and to engage with the deeper ethical debates behind his thinking, he was justifying not only a conversion but a loss—of his former identity as a pig farmer; as an omnivore; as a person validated, even romanticized, by the larger society. To give Bob credit, he was working hard to avoid the mantle of victimhood. But it was very clear that the story of who he was had changed and that a part of his existing psychological identity, of who he knew himself to be, was about to die.

Even if the death of a part of our identity is self-inflicted, it nonetheless remains painful. What I saw in Bob's rigid, masculine movements, the tightness of his T-shirt tucked into his trousers—a shirt he kept on tucking back in whenever it became loose—was an attempt to *hold up* and restrain the image of himself as a "normal man" in a "normal" meat-eating culture. (I thought of my own saggy jeans, and the belt in the back of my car, and how, for me, arriving at Farm Sanctuary was about loosening myself up, of letting down my guard, of becoming unbuckled.)

What Bob was trying to grapple with, it seemed to me, was not his image of himself as a slaveholder or murderer but a sense that his fellowship with "normal" men was over, and that they would reject him. Bob was experiencing what J. M. Coetzee might call a "disgrace" in the eyes of his fellow men. It is this shame that perhaps motivated Bob to remain open to helping other pig farmers improve their systems of farming and slaughter. (In his notice that his pig farm was officially closed, Bob still offered his consultancy services to others.)

I could see clearly that Bob *needed* Gene to be present for his story. In Gene, Bob saw a fit, healthy, and active individual; a model of the successful, masculine male who was also a compassionate, animal-loving advocate and activist of considerable stature and accomplishment. It was evident to me that Bob needed Gene not to discount his (Bob's) previous successes, which brought him comfort. Like my belt that I would not throw away, Bob still needed the notches that validated his achievements in a predominantly male world, made through the domination of the bodies of others—in this case, pigs.

Bob's story is being made into a film called *The Last Pig*.[23] The name troubles me a little—is Bob the "last pig" freaking out as he realizes he is now alone? Does this deny and lose sight of the two thousand pigs he has sent to slaughter? My discomfort comes from the fact that, having made the decision to stop pig farming, Bob still sends his final pigs to slaughter. But perhaps this is unfair—most of us who have found our way to animal advocacy have also come through our own exploitation of animals. In the trailer for the film, Bob talks about how he, too, is "haunted by the ghosts" of those he has slaughtered. At one point, he looks into the eyes of one of his pigs and shows us that he too has been "seen seen." For Bob, the lingering disquiet of that pig's gaze must be, I imagine, deeply troubling.

The new pigs from Bob's farm would be a problem, Susie told me. The sanctuary in Watkins Glen was already home for a dozen or so pigs, and pigs not only take up a lot of space but are very hierarchical, as Orwell's *Animal Farm* got right, and new pigs coming into a community can cause havoc to an established order. Susie wasn't sure what she was going to do or where they were going to go. But their arrival was still a few days away and she had a little time to plan for how they were going to meet the new pigs' wants and needs.

In her book *Animal Capital*, Nicole Shukin argues that standard farming practices meet only the very basic needs of animals, and these only to satisfy profit motives, and *never* the animals' "wants." Intensive processing rests upon what she calls a "breaking" and a "denial of 'becomings,'" (31) where farmed animals are kept in a "limbo economy of interminable survival" where "coping is all" (39). Sanctuaries, when run well, are the opposite. They provide for the needs *and* wants of the animals through the co-creation of atmospheres of respect, consideration, and fellow-feeling.

As I discussed in Part I, trauma, whether mental or physical, expresses itself through the body. Trauma occurs when the sensory inputs into our body–minds—from inside and outside—overwhelm us. We must either mobilize in response (run, fight, protest, act) or have somewhere safe to go or someone whom we can call to our aid. This is a caregiver, and it might be a parent, a partner, a close friend. The latter option may apply to animals, as well: as geographer Henry Buller says, what animals might want is "to be with each other even, in some cases, at the moment of their ending" (167). This is certainly the case for the last pig in the pen at the slaughterhouse. Relief from trauma, in whatever form it arrives, is sanctuary.

When we have nowhere to run to, and no ability to move, our senses shut down. When this happens, our neurological processes are, as Besser van der Kolk puts it, "knocked out." His work has been to reignite those processes that would allow victims to begin to *sense* their bodies again, as a first step. Not only is his trauma center in Boston a sanctuary, but it aims to make the sufferer's body a sanctuary again, too. When they first arrive, many of van der Kolk's patients have little or no proprioception. They cannot feel where on their body the masseuse is touching them. Slowly, they are brought back to their bodies as a safe space. Through breathing techniques, mindfulness, yoga, massage, and other physical therapies, they restart in their minds—their embodied minds—the processes that got "knocked out" by the trauma. They begin to *sense* again and thereby *make sense* of their bodily experiences.

Isn't this what Susie and the caregivers at sanctuaries do? Isn't this what each person who cares for a traumatized animal does? Nonhuman animals in the food system are traumatized bodies without exception, regardless of the label ("free range," "humane," etc.). Trauma is the condition of all but the exceptional few who arrive at sanctuaries, and the people who bring them into the fold, give them space and understanding and respect, and provide the love and compassion they deserve, deal with this trauma by offering their own caring bodies as sanctuary.

What sanctuaries in both these senses offer is the chance for new narratives to develop in the brain–body–world entanglements of those who have suffered. The words we use to describe other animals change within the sanctuary setting: she or he has a gender; she or he is some*one* not some*thing*; she or he is a *who* not a *what*. Trauma is released through body-to-body contact and caregiving, trust, and acceptance—both ways. Providing the individuals—named, noticed, given particular care as a being in her- or himself—a chance to put the traumatic incident into its place in the narrative of their lives is central to that recovery. This brings about true physical change.

For Pigs and Humans . . .

The stories of vegan journeys are often made up of traumatic experiences, and for some people the journey seems too painful to take. So not only do we need to think about the psychology of identity in changing the body–minds of those who continue to exploit animals, but as a movement we need to provide the ability for people to physically mobilize their fear and anger, and also provide a safe place, or safe person, to run to. We need to offer an unbroken narrative of compassion and care. From here the narrative of who most of us were—exploiters of animals—recedes. But it does not, nor should it, disappear completely. It becomes part of our story, and at best it can be a source of knowledge and empathy and an engine for compassion toward those who still exploit animals. Histories that negate our own pasts can too easily become justifications for intolerance or righteousness. Rather, narratives of how people change help us communicate to others that we can all act creatively and imagine living differently—activists and pig farmers alike.

Photojournalist Jo-Anne McArthur regularly visits Farm Sanctuary. It is her "happy place" where her own trauma—her diagnosis of PTSD for what she has encountered in the vast matrix of animal exploitation—can be undone, re-narrativized, healed. It is through bodily encounters with nonhuman others in homes and habitats where we are, optimally, their equals, that this healing takes place. Sanctuary transforms what we think about farmed animals and how we feel about ourselves—forever changing us through different food choices, lifestyle practices, and advocacy.

6

The Save Movement and Bearing Witness

WE COUNT FOURTEEN trucks this morning. Anita Krajnc of Toronto Pig Save keeps a record of every vigil in her notebook. There are two hundred pigs in each truck. Those bearing witness know that Fearman's Pork Inc. slaughters up to ten thousand pigs a day. *Fearman's!* The pigs are transported from all over Ontario, all at roughly the same age and weight.

The first truck is driven by a woman with blond hair and sunglasses. Anita and Hannah have seen her before. Sometimes she stops at the entrance to the plant so volunteers can hurry to the sides of the truck and through the air holes give the pigs watermelon or just a few drops of water or a pat on the snout. Or an apology. Anita and Hannah discuss whether the driver might be sympathetic, since her behavior contrasts with that of the drivers of the other trucks: men wearing black T-shirts, beards, and caps, who wheel around the corners without slowing. The Pig Save activists take a real risk of being hit by the edges of these vast metal containers to get their photos and their last touches of the pigs who in only a few minutes will be slaughtered.

You can hear the animals' screams over the cacophony of hydraulic brakes and diesel engines. Anita points to a huge white tank of gas in the slaughter plant. Written on the side is the label: CO_2. At this scale of operation, the pigs need to be gassed before having their throats cut. The chambers "utilize the natural curiosity of the pigs" (according to the manufacturer's website) to make them enter more pliantly.[24] Advocates for CO_2 chambers argue they offer a more "humane" and efficient method of "stunning," although footage from Animals Australia—the first to come from inside such chambers—shows the pigs in severe distress.[25] Pigs struggle to breathe for around thirty seconds before they fall unconscious and are then dragged out onto the slaughter line for their throats to be cut. At least they're all together, I think. There is no last pig. The CO_2 is a tasteless and odorless irony: we kill them now by subjecting them to our greenhouse gas emissions, literally choking them to death in our waste air.

It is part and parcel of the contemporary rhetoric of industrial animal agriculture that the language of efficiency and productivity is striated with that of welfare and sustainability. In recent comparisons of pig-raising, indoor production has been touted as the most beneficial and practical, assuming "both indoor and outdoor systems are well-designed and operated" (403). Animal scientist John J. McGlone claims that we have only two choices "as we develop more sustainable systems; either change the systems we have or build new systems" (404). For McGlone, gassing ten thousand pigs a day is a positive step in the sustainability of pork production, because not only *must* we eat pigs but we also much find ways to *kill more* of them. The option of not doing either is inadmissible for this man funded by the U.S. National Pork Board.

It's true that a swift death would be a mercy for these pigs, who will most likely have spent the entirety of their short lives—around six months to "finishing" weight—without breathing fresh air before they are crammed onto the trucks. During the journey many animals will collapse on the floor of the trucks, dead or dying. At the height of summer, Save members have seen the pigs frothing at the mouth because of the heat; in winter their bodies may be frozen to the metal sides of the trucks. Whatever the season, their bodies are often bruised and bitten, scratched and covered in feces. Unlike Babe, these pigs don't have the option of an open-topped truck; no wonder she took a leap for her life.

Standing alongside the volunteers as the trucks roll up, I can see that there are as many different emotions in the pigs' eyes as there are eyes looking at us. Some are numb or blank or without hope. Some seem to be dissociating themselves from the conditions. Unbelievably, some still appear hopeful and inquisitive and forgiving, as the animals shuffle over, take our water and watermelon, and sniff our fingers with the snouts they poke through the air vents. I like to think they welcome us for our witnessing. They are advocates all, calling us to their aid, and they suffer our hearts. That is, witnessing is painful, but the pigs *suffer* us in the other sense of the word, too—they bear us; seeing them makes it possible for us to continue with our advocacy, to make it bearable.

Yet we leave them as the trucks pull away as the lights change. We walk away as their advocates. We call others to their aid—the people who pass by in pickups and cars: some ignore us; others "honk for humanity" and wave; some take our leaflets and literature; others roll up their windows as we get close and turn up their radios so they cannot hear us ask them to think of the pigs.

After the first truck leaves, I look at the photographs I've taken through the holes in the side of the truck. One pig's eye has a ring of blood around the black pupil. The eye does not, however, look at me. Instead, it focuses to his left, as if the eye is unable to bear me bearing witness to his suffering—to be "seen seen."

Of course, in suggesting the above, I'm opening myself up to accusations of anthropomorphism. But I'm not going to balk at using such formulations, because an empathic anthropomorphism can help us change our relationship to other species, as Lorraine Daston and Gregg Mitman argue in *Thinking with Animals.* This anthropomorphism isn't the one criticized by "objective" scientists, but rather an understanding that our forms of thinking are on a continuum with nonhuman forms of thought—for instance, terror, grief, bewilderment. I detect a sad bewilderment in the eye of the pig, almost shame in his powerlessness, a wish that his creaturely exposure be taken away—as if *he* has done something wrong; as if the shame is *his* and not mine. I sense that, if it could, his spirit would levitate out of this hog truck in the way Babe fought to alight over the backs of others, detach from the fate that is destined for their bodies.

This is not to render animals as absolute victims; or is it? Would that yield to the truth? As Anat Pick argues, to "speak of an animal's vulnerability in this context is to draw attention to their outstanding position in the judicial, political, and moral orders" (15). The animal is first rendered powerless so she can be rendered into constituent parts for the market. There is nothing to give back to the creature, no revenant, not even her rind. Nothing to witness here.

When the truck has gone and with it the pigs, three of us remain on the small island in the middle of the street by the traffic lights. I don't quite cry. What kind of tears would I shed? Grief at what I see? Frustration at my powerlessness? Relief at doing *something*, at least? A mixture of all three?

In fact, I don't quite know what to do. I want to imagine my way into the mind of that pig who would not look me in the eye, yet worry that if I do I'll break down. Hannah is tearful. Christine, a teacher at a school down the road, runs over to get more leaflets. I stand between the two of them not knowing how to think. I'm not quite ready to leaflet. Like many new activists, I am anxious about confronting others with my beliefs, especially those who may not feel the same way. Because I don't know what to think about what I'm seeing, I can't draw myself into conversation with passersby. I can't, if I'm telling the truth, look *them* in the eye. I know, deep down, that I'm ashamed: of these drivers *seeing me*; of not yet fully bearing witness; of not yet letting go of the need and desire to be part of the "normal," meat-eating, animal-exploiting world; of needing *to be like* everyone else and liked *by* them. I have not yet the bravery or reassurances that everything will be okay, to fully let that identity go.

Yet here I am, bearing witness. And being witnessed is a little more bearable than before.

The History of the Save Movement

I'd immersed myself in the videos from the Toronto Pig Save slaughterhouse vigils before I came across the image of Babe making her escape. The Save Movement began when Anita adopted her beagle-whippet mix Mr. Bean. Anita was already a feminist and activist, having been trained as an instructor in social movement tactics and lecturing weekly at Queen's University in Toronto on the subjects. Anita is a scholar of Tolstoy and Gandhi, and much of her community organizing is inspired by their activism, driven by social justice and equality through an approach that she says is based in love. At the heart of her work is active witnessing.

Anita knew about the Quality Meat Packers pig slaughterhouse near where she lived in downtown Toronto. When she and Mr. Bean took their morning walks each day along Lake Shore Boulevard she found herself face-to-face with trucks packed with pigs. She could see their faces and their sadness through the air vents. Coming up close to the pigs every day forced her into action.

Beginning in December 2010, Toronto Pig Save held monthly vegan potluck meetings to raise funds and organize. The mission was to metaphorically give slaughterhouses glass walls by showing those outside what went on inside. The activists collected donated art, photographs, and illustrations that depicted the slaughter of pigs. They held art shows and built an online gallery. In July 2011, they began the weekly vigils. They printed leaflets and drew placards and stood where the trucks passed every day and were sometimes forced to stop at long red lights. The activists started to give the pigs water, watermelon, and take photographs. They began to speak to the drivers of the cars idling at those red lights. Now, the Toronto groups (which are Pig Save, Cow Save, and Chicken Save) organize three vigils or leafleting events a week. The Save Movement has spread to over twenty countries, and is related to movements for bearing witness in Israel, Germany, Spain, Portugal, the Netherlands, and the United Kingdom.

"Bearing witness is the most important thing we do," Anita told me. "Being physically present to witness the injustice faced by farmed animals throughout their lives enables each of us to more fully empathize with their plight. . . . It's only since I joined together with others to collectively bear witness to the cruelty of slaughterhouses that my level of commitment increased, so much so that it has become a lifetime priority for me."

Melanie Joy, writing in *Why We Love Dogs, Eat Pigs, and Wear Cows*, agrees: "[W]hen we bear witness, we are not merely acting as observers; we emotionally connect with the experience of those we are witnessing" (138). She continues: "[C]ollective witnessing closes the gap in social consciousness" (139). Collective witnessing is about being "seen seen"—being seen by the animals, by the passersby, and by other activists who support and validate the witnessing. If dissociation is the feeling of not being fully present or conscious, then active witnessing, especially in a social collective setting, is the opposite: a sense of being present and conscious to what is happening before us. And as I've already argued, witnessing is not merely a cognitive or perceptual act. It is a felt experience, a somatic, corporeal entraining. It is, or at least can be, a "tragic spectatorship" centered on feelings of empathy. Empathy, in turn, is the way our brains fire so that we believe and feel what the other is experiencing. When we witness the pig's shame, fear, and bewilderment, we can experience these, too, with the right moral attention. We experience them in the place that keeps score of such things: the body. It is why active witnessing of the actual animal bodies, whether in motion or immobile, is central to the sense of empowerment that activists from the Save Movement say they experience.

Seeing the pigs in their powerlessness would make us feel powerless, save for our ability to move our own bodies, to push the traumatic experience into a narrative. This is why activism is about *movement* and *mobilization*. Both these elements in countering trauma are central to active witnessing. As we will see in the next chapter, this insight prompted a question: How could I move my body in such a way that not only empathized with the pigs' powerlessness but also converted our common vulnerability into action on their behalf?

Making Mischief

On the island between the flows of traffic I talk for a while to Hannah. She is a student of Buddhism and is learning to practice *ahimsa* ("nonviolence"). She tells me how much she admires Anita's work in establishing the Save Movement. I note the gender makeup of the regular

Save events: mainly women. It would be good for me, as a man, to take the leaflets and approach the cars. Hannah is fearless in approaching people and handing over literature, even when they want to disengage. Compassion is her driving motive—for the pigs and the people in their cars. For Hannah, as for Anita and others, bearing witness to the endless slaughter of nonhuman animals is their calling. They have been touched by the bodies of pigs for longer than I have. I'm still learning.

We literally touch the pigs, too. Seeing them in the trucks, tickling their snouts, not knowing, as a novice, if they would try to bite me or not, I feel their fear and, yes, the very curiosity that is instrumental in the death they will shortly undergo. Psychologist Todd Kashdan at George Mason University has shown how curiosity and anxiety are incompatible emotions. Turn one dial up, and the other goes down. I hope that this is one thing we can do for the pigs: that our touches alleviate their anxiety, if only for a moment. As it turns out, it's our touches that get us into trouble.

A little while into the vigil, a police patrol car pulls up and the officer asks Hannah, Christine, and me to go over to the sidewalk. The officer does a U-turn around the lights and pulls into the entrance of a Tim Horton's. She gets out of the car and walks over to us. She tells us she is responding to two phone calls, which she doesn't think came from the slaughterhouse. (We learn later that one of them did.) The officer had seen us go into the road and touch the trucks and this could be considered an offense.

Anita argues back. We weren't obstructing the trucks, she says. We were trying to give water to the pigs, and some comfort. Did the police officer know what condition the pigs were in? We'd seen pigs terrified, scratched, and beaten, and Hannah had seen pigs dead or dying. It was against federal law to transport animals for over thirty-six hours without water. *Thirty-six!* Was the officer going to investigate the dying pigs?

"If you're touching the truck then the driver has to be aware of you, he can't risk hitting you," says the officer, "and if you're doing that when you're standing in the road then that's an offense."

"What offense?" asks Anita.

"I don't know, but I'd guess mischief," the officer replies.

Anita responds by asking the officer to take some of our literature, to come and bear witness when the next truck arrives. The officer isn't having any of it.

"Now I'm not questioning your right to protest, ma'am," she says. In fairness, the officer isn't looking for a standoff. In fact, she gives me the impression of being genuinely concerned for our safety. "I watched you," she points at Hannah, "and you," she indicates me, "and I saw you step out into the road to touch the trucks. That can be seen as an attempt to obstruct, and that means I'm going to caution you for mischief."

Anita has been in this position before. She keeps pressing for the officer to acknowledge our reason for being there (to alleviate the suffering of the pigs and to bear witness to their vulnerability) and our means (love-based compassion, nonviolent and law-abiding action). She asks again if the officer wants to take a leaflet.

"I know what's in the leaflets," she says, pulling out her notebook.

"Then you know how much they suffer," Anita replies.

The officer looks at a loss. It is the moment when her identity—as the enforcement of not only traffic law but a law of human exceptionalism, which says that stopping a truck in its task of delivering animals to slaughter is a crime—unravels. What is the real crime? Where is the genuine mischief? What passes across her face is acknowledgment that the difference between us is in perception only. Then the moment is gone, and she is asking for our names.

"I'm heading home soon," I say. "I won't cause you any more trouble."

The officer stares at me. Who is this British man in a group of Canadian women in a protest against the slaughter of pigs?

She does not press for my name. I'm relieved; I don't want to get in trouble at passport control. But I'm also glad to witness this engagement between the activists and the law and having it clear on which side the law operates. Bearing witness is not only about "cultivating integration" between one's values and one's acts, as Melanie Joy expresses it. Bearing witness is about being present at conflicts where the truth of a situation requires people to be present, to hold power to account.

Witnessing as Saying the Unspeakable

Bearing witness has a long and powerful history—back perhaps to Nussbaum's "tragic spectatorship" in the theaters of Ancient Greece. More recently, bearing witness has been fundamentally altered by the last century's human atrocities, especially the Holocaust of World War II. Many philosophers have thought deeply about the moral aims of bearing witness for human victims of injustice. W. James Booth writes that those bearing witness become living reminders of a past and serve as sentinels, urging others to remember, too. To bear witness, says Jeffrey Blustein, is to assert the moral status of victims and their equal membership in a moral community, exercised by giving them a voice.

Bearing witness to the suffering of nonhuman animals, however, exposes the existing entanglements between humans and nonhumans. As an act of *witnessing*, it reveals our means of perception and, importantly, the way we think about how we perceive others. To consider the animal him- or herself as a participant in the witnessing is a powerful means of shifting those boundaries. We can also go further in questioning the nature of perception itself. How we perceive others shapes what we think and feel about them, as individuals and as groups. Perception is generally thought of as a process of thought. And yet perception is not only this: it is, as I hope I've made clear throughout this book, a physical act. Indeed, neuroscience has shown that our perceptual systems are the basis for our conceptual systems—which are then fundamentally phenomenological. What if we perceive *perception* differently—as enacted within our bodies first of all? Such a recalibration of *how* we perceive will change *what* we perceive.

Bearing witness is in this sense a physical process of perceiving with moral attention. As Melanie Joy puts it: "Virtually every atrocity in the history of humankind was enabled by a populace that turned away from a reality that seemed too painful to face, while virtually every revolution for peace and justice has been made possible by a group of people who chose to bear witness and demanded that others bear witness as well" (139). Active witnessing moves the body (notably, the face) toward compassion. Its opposite is disconnection. When we look away we disconnect from the suffering of others. We do this in the physical act of not looking. Just as witnessing moves our body in a certain way (opened up, often standing), looking away moves us into closed positions. Our body keeps the score.

The Save Movement and other groups make witnessing not only essential to the remembering (itself an embodied expression of re-membering) of tragedy but removes the blinkers from our view of the world.

Witnessing can stimulate solidarity and the strength gained by acting together in a common cause. Yet it can also untap a wellspring that can flood our bodies with grief at the unspeakable.

One evocative example of this complex reaction to witnessing are the silent vigils organized (first of all) by the group Animal Equality,[26] and later as well by Our World, Theirs Too, which in 2011 established in North America a National Animal Rights Day.[27] Activists are arranged in orderly rows in a public space such as a city center or square. Each activist stands in silence, holding in their arms the body of a dead nonhuman animal while a single speaker addresses the activists and the crowds who gather or who are already present (such as those sitting at restaurant or café tables). That speaker tells the stories of the lives and deaths of the particular animals, or, if that information is not available, of the brief existences of animals just like them.

After speaking to those who have attended these events and watching videos of them online it seems clear to me that the most persuasive element of these vigil protests is the presence of, and the relation between, the bodies of the activists and the dead nonhuman bodies they carry in their arms. Many activists openly weep. Those who don't wish to hold an animal, or when the animal her- or himself is too large, lift up a picture of that animal. When at the end the nonhuman bodies are gathered again to be taken away to be cremated, some activists cannot let go. Afterward, many seek solace and affection from one another to remind them that their feelings and reactions are valid and have been seen.

The orderliness, the dignified silence punctuated by only a poem or song, the solicitude with which the animals' bodies are treated at every stage of the vigil—the opposite of how those animals may have been handled when they were alive—give this witnessing ceremonial weight. By focusing the attention of everyone present on the animal held in the arms—a kitten, a goose, a faun, a lamb, a chicken, a rabbit, a piglet, a foal, a raccoon, a fox, a rat—the vulnerability of every body present is honored: the one who cradles as well as the one who lies in the arms.

After the vigil may come the speeches, with every organization director needing to have their say, and the power is diminished, as we are taken out of the embodied consciousness of the animals' deaths and our memorial obligations to them and placed back into the non-corporeal world of words. What's so striking about these events is how, at their most impactful, they allow bodily knowledge to shimmer with affect, to make abundantly clear that brain–body–world entanglements exist across species boundaries. Perhaps we'd recognize this more often if we gave the time, space, and silence for the unspeakable, the unbearable, to carry itself into meaning and the silent body to bear what formerly could not, but now can and must, be borne. Remaining open enough to such a burden—the *bearing* of witness—can be laborious.

However, as Kathie Jenni, Professor of Philosophy and Director of Human–Animal Studies at the University of Redlands, California, has written: "How to bear witness is a matter of moral judgment that those who would honor the animal dead must take on. That struggle for wise judgment is itself a labor of respect."

Back to the Vigil

The officer drove off after warning us not to step back into the road. It was almost 11 A.M., and we had a train to catch. We packed away the placards and said our farewells. As we walked, Hannah, Anita, and I passed the entrance to Fearman's at the same time as another truck arrived. The light that allowed the truck to turn into the processing plant (the civic street facilities structured to make the slaughter of pigs easier) was red. Hannah and Anita got their cameras out and I took the decision to run on ahead. I stepped in the road as the light went green. I walked as slowly as I could to obstruct the truck. The driver scowled at me, pushing and jerking his rig. I bent down and retied my shoelace, which offended him mightily, since I was in the middle of the road. He edged right up to me and I moved as he accelerated. The angle of the turning truck pushed Anita and Hannah back so they weren't hit or hurt. And all the time the pigs were ever closer to slaughter.

The activists who organize and show up at the vigils I've mentioned in this chapter are vital witnesses. But no less significant are the photographs and videos gathered from these encounters. Circulated rapidly across social media, these images have the ability to reach thousands, even millions of people; indeed, they brought me to Toronto. When people watch this material, are they also bearing witness? Yes. But watching video does not have the same impact as the active witnessing of being present and putting one's body into action where animals are suffering: even if only to hold a creature in your arms; even for the time a truck sits on red.

As with the image of Babe, the shame of the pig who would not look me in the eye has stayed with me. I'm glad I made mischief on his behalf. And there is a trace of mischief in the very word itself, which comes from around 1300, when it first defined an "evil condition, misfortune, need." It derives from the Old French *meschief/méchef*, "to come or bring to grief, be unfortunate"—the opposite of "achieve." At its root is the Latin *caput*, which means "head." Mischief, then, emerges from the same source as the word we use for cattle.

What were we doing that morning if we were not, quite literally, "coming to grief"? *Come to see* the grief of the pigs. We came with our physical bodies to a place of grief and grieved as we gave ourselves to witness this "evil condition, misfortune, need" of the nonhuman animals trapped in their absolute vulnerability. In doing so, in being literally cautioned for mischief, hadn't we actually, in that cautionary moment, achieved our mischief, the physical movement of *coming* to grief, and making visible the invisible "evil condition, misfortune, need"?

"The paradox that works in our favor is that bearing witness makes activists stronger, not weaker," said Anita on the train home. "We can sense that collectively we're making a difference. We help the public make the connection with the trapped pigs in the transport trucks and the food on their plates. I and so many other activists feel stronger and more determined than ever because we have a course of action we feel makes a difference."

This *feeling*, it seemed obvious to me, was both noun and verb.

—

Anita and the Save Movement, the work of Animal Equality, and other groups such as 269 Life[28] (originally founded in Israel, they have enacted the branding of animals with hot irons on their own flesh) are examples of where campaigns to be compassionate toward both humans and nonhumans take shape, shockingly and disruptively, around the body. Such mischief making would be at the heart of the next embodied action I attended the morning after Toronto Pig Save. I would put my body on the line to mobilize my anger at the exploitation of animals by running forty-four miles.

7

The Slaughterhouse 66k

BEFORE THE ARRIVAL of rail and shipping in the nineteenth century, the taking of farmed animals from summer to winter pasture or from farm to market and slaughter often took days, sometimes months. These droves are recorded in place names, pathways, and rest stops throughout the United Kingdom dating back to Roman times. The passage of nonhumans was impossible to ignore as huge numbers passed by. In the seventeenth century, Daniel Defoe recorded that 150,000 turkeys were droved from East Anglia to London each year, a journey that took three months to complete and must have offered up remarkable sights, sounds, and smells to those who encountered it. Occasionally, as in the case of several years ago when protestors attempted to stop the live export of animals across the English Channel, these journeys become visible and intolerable to (some of) us.

As I write, thousands of refugees are being transported toward Europe on boats, trucks, and trains, or are walking on foot, to escape the conflict in Syria, civil war in Libya, and instability and insurrections across the Middle East, Afghanistan, Pakistan, and Africa. Some die through drowning on flimsy rafts or suffocating in unventilated vans. It is hard not to see in the leap of Babe into freedom and the visible refugee crisis affecting Europe a collective rebuke to our refusal to notice the numerous invisible and unrecorded victims of our trafficking in flesh, and who pass by without us caring.

A surprising example of the power of bearing witness to effect a transmutation of subjectivity between human and nonhuman took place in Australia, where two concurrent media stories—concerning the live export of animals abroad for slaughter and the rise in asylum seekers reaching Australian shores—exposed the entanglements already there. A part of the process of seeking asylum is for refugees to provide testimony and bear witness to their own suffering. However, the Australian media and many powerful commentators had already dehumanized the asylum seekers as figures of hate through, as Gillian Whitlock records in her study of the events, stereotyping and repetition. However, two graphic documentaries were aired on Australian television around the same time: one documenting the journey that the asylum seekers had taken and another the conditions of the live export of Australian animals to Indonesian slaughterhouses, where the animals were shown to be brutally abused and tortured. The Australian government halted live exports pending more "humane" conditions being established. The graphic imagery of the documentary bore witness to the treatment of the animal, and the nonhuman became, for a moment, as Whitlock writes, "an ethical subject."

Activists began to use social media, especially Twitter, to communicate the benefits gained for the animals to the aid of those seeking asylum. They began sharing messages that talked of the "live export of refugees." It was then, says Whitlock, "through testimony of the suffering of animals in the context of debates about public policy that the alterity threshold in perceptions of refugees in the public sphere was challenged, and asylum seekers were brought into that category of creatures to which we attach rights and feel obligated" (92).

The image of Babe's escape threw up a similar conflation of categories that, I felt, opened up a possible response. I had an idea to find a factory farm or stockyard and trace the line that the trucks take to the slaughterhouse. Before I arrived, Anita and another activist, Susan, located a stockyard forty-four miles outside Toronto, in Cookstown. We called the event the "Slaughterhouse 66k" and it was catchy enough to attract the attention of a Canadian TV crew. The run would follow a route from the stockyard to the Ryding Regency and St. Helen's slaughterhouse and meatpackers. At the end of the run,

Anita and others would hold a special vigil. The hope was that this run would make a statement and draw a line—a long line, upon which I would put my body, to make visible the invisible route these animals take on their way to slaughter. But not only seeable. Something embodied. *Feelable.*

Entangling Running and Animal Advocacy

I've been a runner since I was sixteen. Before that, I loathed running, although I enjoyed other sports, especially soccer, rugby, and cricket. Running seemed like eternal punishment. My freedom to move at will was taken away in long cross-country races. Every time I stopped, my gym teacher would shout at me to get going again. Then something changed: running became a mode of transport. As friends began to leave home and move into their own apartments, and since passing my driving test was still a little way off, running became a way of getting around my neighborhood. Beyond the practical component, running also provided me with freedom from a still difficult home life.

It wasn't until my transformation at thirty-five that I began to take running more seriously. Previously, I'd been plagued by weak calf muscles and a stiff back; I'd undertake a typical training routine for a long run, and then give up for the rest of the year. As I learned to sleep, fixed my relationships, and moved toward veganism, I also joined a running club and started training regularly.

I began to experience and enjoy the sociality of running (indeed, I could talk about running for hours!). Fixing my destructive habits meant that I no longer pushed my body beyond its boundaries. I began to take care of myself and my body began to take care of me. I became a consistently good runner.

I started to wonder how to integrate ideas of movement and freedom into animal protection work. This was a selfish motivation driven by a lack of time for both marathon running and animal advocacy. More philosophically, I wanted to know, as Martin Rowe asks, if "running has the potential to reveal to us the animal that has been hidden in the meat or dairy that we eat" (32). Could my running *be* advocacy?

Then I came across the image of Babe and the videos from the Toronto Pig Save vigils. All those trucks. How far had they traveled? How much farther did they have to go? *And where exactly did they come from?*

On the Run

The morning of the run I woke at 5 A.M. and had a banana and a protein shake. I'd packed my kit the night before: energy bars, water, rehydration salts. I'd been fearing a hot day, but the temperature had cooled and it was expected to be only sixty-five degrees, maybe seventy. Despite some apprehension, I was excited for the challenge and to soon be going home, too: the day after, I had a plane to catch. I was excited most of all to set the plan in motion. I would be running farther than I'd ever before. It felt suitably Churchillian: not the beginning of the end, but the end of my beginning as an animal advocate.

Anita and Paul, another activist, arrived to pick me up and as we drove along the highway the scope of what I was about to do set in. *I was going to run all this way back?!* I took a deep breath: forty-four miles wasn't too far over the course of a day. It wasn't a race. I had support all the way.

We parked in the empty lot next to the stockyard. It was 7 A.M., and Christine and Rob, two fellow activists who lived in Cookstown, were waiting to greet us. We introduced ourselves while another activist, Katie, climbed into her cow-patterned costume and another, Michael, set up his camera on its tripod. Anita unfurled the banners to get some initial shots to share online in motivating tweets. The plan was to play "Amazing Grace" over a bullhorn from Katie's iPod, and then I'd set off.

The stockyard was already busy, with trucks and trailers pulling in, some with multiple animals for sale. As we started the music, the owner of the stockyard, a woman in her sixties, came toward us accompanied by a man.

"You people are sick!" she screamed.

A shouting match ensued, regretted afterward—compassionate behavior is a better way of expressing advocacy. The argument was surreal. The stockyard owner's attacks were vitriolic: that we'd rather the animals were abandoned to die alone than have the "kindness" of being slaughtered; that it was because we had the education and cultural resources to know how to cook that we had the wherewithal to also become terrorists! The owner's defense of her business sounded just like what pattrice jones identifies as the "absolute real threat" that meat eaters, especially those involved in the killing of animals, feel in response to activist protests. This woman was acting as if we were threatening her very (way of) life. I saw the altercation as another example of how people respond when they feel the continuity of their identity principles is under threat.

Finally, the owner and her bodyguard left, and we continued playing "Amazing Grace." I noted how fast the trucks were moving along the road and how narrow the shoulder was. After a warm up and a quick pee behind one of the trucks (a wry smile at that), I was off—up a hill, with a folded bit of paper in my pocket containing the directions: Highway 81 to 27, all the way to Islington Avenue, then Albion, then Weston. A surprisingly straight line.

After the first six miles, a car pulled up alongside containing Christine and Rob and their children. The Ontario license plate read GO VEGAN. "I was amazed it was still free," Christine told me about the name, as we stopped and bore witness outside two massive dairy farms. I chomped on an energy bar, and after saying goodbye moved on. I would not see anyone again now until Kleinburg, the lunchtime stop.

No one human, anyway. A few miles farther on, I came across a little tortoise attempting to cross the road. Of all the critters to come across on a long run: a tortoise! I took him a hundred yards from the road and pointed him in a safer direction. He was a reminder of not only why I was doing the run but also how to do it. I was thinking about the time I might run. I wanted to fully engage with the action, to "leave everything on the road." But I was not a hare; this was not a race. More important was that no matter how fast or slow I ran, this act was my choice, based on my bodily autonomy that remained free throughout. Of course, in a human-centered world there was no way I could ever experience the *absolute* vulnerability of an animal within our food system. I am fortunate enough to live in a society where I'm afforded freedom from immobility and subjugation. For the nonhuman, especially the farmed animal, her immobility will not end, in a world that disavows and refuses to respect her bodily knowledges and desires.

And yet. We protest, organize, march. We advocate. We run to and for them. We bear witness to their last moments. If these acts do anything, what they do is *embody hope.*

Being an "Engaged Running Vegan"

The affinity between long-distance running and plant-based nutrition is founded on a realization that predominantly meat-protein diets are not as beneficial—especially for recovery times—as plant-based food plans. The collection *Running, Eating, Thinking* has made further connections between bodily movement and animal advocacy. Some contributors draw a parallel between running as a physical mobilization that expresses our (unfortunately, not universal) freedom to move our bodies as we choose and the widespread confinement of animals who are constrained to the point of insanity. As Gene Baur wrote in that anthology: "Factory farming and the culture that supports it denies other animals the simple pleasures of moving their bodies in a way that they want to. These creatures may have lived every moment of their short lives in a barren cage no bigger than their bodies, but when given the opportunity to express who they are, even the most genetically manipulated of them want *to try* to move" (42) [emphasis in original].

For Catherine Berlot, on her regular runs past a pig farm reeking of ammonia, the task is "to evolve into an 'engaged running vegan.' I am not just running for myself. For me, this means learning to find ways to be an animal advocate while I'm running" (171). There are countless examples of running inspiring and changing others. Its impact is perhaps, as the philosopher-runner Mark Reynolds argues in *Running with the Pack*, due to running's intrinsic nature. We run to run, and it touches people in a way that extrinsic motivators never can. Could this be put to use in leading people to animal advocacy? I knew there would be an idea that could connect animal protection work to running. All those lines drawn on Strava and Garmin Connect could be put to use in making visible the invisible of what happens to animals in our food system. That is something I wanted to explore.

Back on the Run

Twenty-six miles into the run, I reached Kleinburg. I saw a Starbucks. I only had a mile to go to the rendezvous but the sun had come out so I went in, grabbed a water from the fridge, and then waited as *all four staff* helped one woman decide if she wanted a chicken or tuna sandwich. I slammed the bottle on the counter.

"Can I pay for this?" I asked.

I looked unkempt, even creepy, covered in salt and spit. But four people to help one person choose which dead animal to chow down on? I paid my two bucks and headed off.

I regretted being angry. But I'd just done the bulk of the run, a marathon in three hours and fifty-two minutes, and my self-control was not what it should have been.

At our meeting point outside the McMichael Canadian Art Gallery, my crew of three—Anita, Paul, and Katie—were waving banners and cheering me in. They were joined by a couple who'd heard about the event and had driven up from Toronto to add their support, as had a guy from Hamilton, an hour away. Lunch was water, a banana, a Vega bar, and some magnesium and sodium salts. Anita and Katie handed out leaflets to passersby telling them about the run. People took the leaflets. "Good luck!" they shouted as they walked off. I stretched out on the grass, thinking: *Just five more minutes.* Then I lifted myself to my feet and carried on. I could have taken more time; I was well ahead of schedule. Yet I knew if I stopped I'd find it harder to begin again.

As I ran, I thought about the *sense* of running as an embodied experience, how the motor of injustice was a participative element of this run. I'd thought about how this run might go and that it might bring up a physically sensed experience of grief: over the senseless deaths of billions of farmed animals each year. But whenever my mind lingered on these images, I found myself slowing down, my chest tightening, and unable to breathe.

For the act of running, this grief was not useful. For *this* activism, it was counterproductive. To run properly I could not indulge myself in grieving. Instead, I focused on Babe; I ran for the pig who would not look me in the eye; for the cows and chickens I had witnessed as they went to die—but I did not grieve for them as I ran. I ran instead to be of service in making visible their invisibility and making their visibility feelable. It came as no surprise to me that Gene Baur used running as a way to burn off frustration and see things with more equanimity while campaigning to rid Florida of the sow gestation crate. As Baur writes in *Running, Eating, Thinking*: "In running and breathing and exploring the world through the physical body that was given to me, I'm connected to the most fundamental expression of our animal nature" (40). In this awareness of his animality, Baur sees a connection between his own body and the immobility of the animals he works to free. That's why I was running, too.

At thirty-eight miles, I stopped for a bowl of fruit salad. I texted people back in the United Kingdom. I prompted them: *Remember how I'd spoken before the run about long distances leaving space in the mind for only the profound, the truly important? That void that Haruki Murakami identifies in his book* What I Talk About When I Talk About Running. *Remember that? How when I reached that point, the trivialities of existence would drop away?* Yet what I texted my friends was that all I could think about was fruit! This cold cantaloupe melon in my grip!

The break did me good. Back on my feet, I felt light and energized. Glycogen restored, I was looking forward to the rest of the run. I had only five miles left.

Running For and As Animals

If we are to be engaged running vegans, then running has to disrupt our species identity as exceptional humans and self-serving creatures convinced of our superiority over other species. Can it do this?

My belief is: Yes. Running is an embodied act, an embodied knowledge of how to act. Running is one of the bodily movements (along with yoga, meditation, and swimming) that the cultural theorist Ann Cvetkovich identifies as contributing to her "utopia of ordinary habit" that helps her through recurring depression. Running brings Cvetkovich out of her mind and back into her body. It resists the denial of brain–body–world entanglements and, in resisting, helps her remain aware of her relations and responsibilities. Being free means being able to move and express one's natural desires.

Running often provides us with a stark reminder of the corporeal limit of what we *can't* achieve. For this reason, perhaps, Socrates derided bodily existence: it complains, it interrupts, it disturbs, it distracts, it intrudes. Rowe translates this for we runners as "the messy praxis of glycogen depletion, gastric distress, blackened toenails, salt deprivation, bleeding nipples, chafing, cramping, sweat, and spit" (24).

Who are you, though, if you do not move? Trauma takes hold of us when we are immobilized, with no way of moving away from the threat. *Having nowhere to run.* Can we think about this in terms of why people don't move over to "the animal side" or toward climate change?

Turn that on its head: bodily movement stimulates brain movement. Change comes first through embodied action. Combining veganism, animal liberation, and running, like active witnessing, *creates new bodily knowledge*—when it is enacted as part of a new narrative or undertaken with "moral attention" to others unable to move. Might such embodied action offer us another way of peeling back the layers of what separates us as species? As Baur says: "Movement is not merely a function of our being animals, it's an expression of *self-identity*" (43) [emphasis in original]. Can running be seen as an exemplary act of disrupting unhealthy identities, which sometimes lead to the exploitation of others?

Rather than reinforce the autonomous and hierarchical sense of human exceptionalism, running brings us back into our materiality. We acknowledge our right to exist through movement: whether it is instinctual, desired, thought, or thwarted. I run—I push my body—I revel in the freedom and autonomy to move my body, because so many others cannot: refused movement and freedom by us, who can.

The End of the Line

Energy restored, I was able to run the last five miles as if they were my first. In fact, I reached the slaughterhouse finish line too early—the welcome was not ready! I ran a mile down the road and turned and came back again, this time to a crowd cheering me in, snapping photographs. I felt fine; indeed, I wanted to carry on! I met with the visitors, some who were attending their first vigil, some who grabbed banners and stood on the front line. Others wanted to sing and play the guitar. Every now and then I had to go and find a place to be on my own. It felt affecting—as if I was asking for someone to come and ask me how I was. Or perhaps I was just wobbly, depleted of salts. I still haven't made complete sense of the experience—not mentally, anyway. Perhaps my body knows better, and that's enough.

Around seven o'clock I was beginning to feel light-headed. Liz Marshall, director of *The Ghosts in Our Machine*, an elegiac film on captive nonhumans and passionate advocate humans, rounded us up to go to dinner. I was doing my last attempts at a warm down when a truck arrived.

We weren't expecting it. Slaughtering takes place from the morning onward, so most trucks come in the very early hours through to midday. When a truck arrives late, the cows are loaded into a holding pen, where they will stand overnight, confused by their new surroundings, smelling waste and blood. And these creatures want to run. Like Babe, they want to escape their fate.

The activist Mary-chris Staples, with whom I'd taken my first ever step as an animal activist at the chicken vigil in Vancouver, was also present. She followed the truck as it entered the yard.

"You are loved!" she shouted to the cows, as she had to the chickens.

One of the slaughterhouse staff stood with the offloaded cows and scratched their noses. Many workers no longer eat the specific animals they render into parts.

There was so much darkness that I wanted to expose with this act of running, to shine light onto the bodies being broken behind the walls. I wanted to draw a line to point toward it. The vigil was the arrowhead: *Here. Look here.* There was so much about consuming animal products that was wrong, I wanted to tell people. And to hope that things would be better and brighter for all of us. For a moment we felt that hope, myself and those people at the vigil. I'd brought out folks who'd never been to bear witness before. We listened to "Amazing Grace" and sang songs. I was thanked for lifting the mood of a vigil that is normally so mournful. And then we all watched this truck roll in and unload the scared cows.

If I am going to cry, I thought, *it is now.*

But I didn't cry. Instead, I stood up and went to the chain fence. The other activists began a further round of "Amazing Grace." I faced the cows and I started to sing.

—

We have drawn, inscribed, and entrenched the line so deeply between ourselves, as creatures of culture, and the rest of "nature" that we do not fully embody our *sense-making* capacities. The "drawing of a line" is a metaphor, but it has physical properties in the world: walls, laws, fences. The line has expanded ever more in the human species' favor, and in doing so it is shrinking further and further the space and freedom for those on "the animal side" of the cages, moats, and glass. The reinscribing of invisible lines (for instance, the line the pigs and cows travel between a factory farm and a slaughterhouse) is, I hope, a powerful and provocative means of drawing (the verb is instructive) our attention to the lines between ourselves and other species, and how we all live and die. If we lived in balance with the rest of the planet's species we would not have intensive animal agriculture. If we lived with ethical compassion and a respect for the desires of all life, we would not have lines between us at all.

Living in balance with the rest of the planet doesn't mean a return to a putative golden age of symbiosis, respecting the animals we individually kill, as indigenous peoples are often supposed to have done. It means taking into account our sense of what a population of nine billion people will mean for the resources we have already used and those that remain.

What I know now, at least, is that a new line has been traced, a thread in the journey of how a living animal becomes a packaged meat product, traced along a visible surface. It runs from the Ontario Stockyards in Cookstown to the Ryding Regency and St. Helen's meatpackers in north Toronto and is forty-four miles long. It is not on any fancy running website, but it is a line inscribed in the minds of the people who were there and in my body.

I have run half marathons, marathons, and other longer runs for the sheer pleasure and also, sometimes, for the risk and danger—I once got trapped up to my thighs and elbows in mudflats a mile from any human habitat, with the tide coming in. The Slaughterhouse 66k was my first long-distance run for the animals. The run was, as I've hinted at, the end of a longer journey. Reaching the abattoirs in downtown Toronto was an arrival after not only a few hours but five years. I'd arrived in a place, but also with an identity: as an advocate for animals. I knew then this line I'd run was not for me. As the filmmaker Pier Paolo Pasolini once put it, "I want to express myself by throwing my body into the fight." I'd finally found my place on the front line in the war; and it was on "the animal side."

Epilogue

Finding a Voice

THIS BOOK IS a story of change as it was played out through my body. I've explored the relations between bodies as they have run, escaped, witnessed, cried, breathed in the air, and died. Spending time with Babe—her image but not her body—sent me on a journey to learn what it means to be entangled with others in bodily encounters, witnessing their lives and their deaths, experiencing their vulnerability and shame, and my own. I wanted to find my place on the front line of animal advocacy. This line of narrative explored how rediscovering a relationship to my body began an opening-out to a more compassionate, vegan, flourishing life.

So I want to say once again that attending to the body cannot be left out of advocacy work; bodies are at the very center of the potential for radical, global change.

Sometimes this means helping others come to terms with their corporeality or bodies that are not working well in their sense-making capacities. Long-lasting change will come about not because we first changed our minds but because we were able to alter how we sense the world and how we feel for the other bodies with whom we are already entangled.

Since returning to the U. K. after my fellowship, I've taken part in animal activism protests and volunteer on vegan education stalls with my local group and fellow advocates. I have found my place there, and here on the page, as an advocate-writer. The body I have rediscovered is not only physical but also an entry to the body of work in the growing corpus of knowledge that is Human–Animal Studies, to which I hope this story contributes.

My journey has brought me to the central question of our time. The aim of animal advocacy is to bring about a change in our species' relationship to others and mitigate the effects of climate change for our mutual benefit. First and foremost, we can all change what we eat. Vegan lifestyle practices are not only compassionate toward nonhuman animals but also essential in building a sustainable global future. But going vegan is not the *last* thing we must do. To change how we relate to other species, advocacy and activism are our next physical steps.

What does it mean to be an advocate? The Latin root of the word is *advocare*, "to call to one's aid." To be an advocate for animals is to call our fellow human beings to the aid of nonhuman animals in their suffering.

But there is another meaning in the term. If the word literally means "call to one's aid," then what we are doing is also calling *to* the nonhuman animal—to Babe—for *her* aid. Our supposed greater intelligence, complexities of language and social systems, and our mastery and advancements have not led us on a path of sustainability and security. We can make mischief all right, but, as Susie Coston lamented, it often makes no sense. Although we have overcome a great swath of environmental and pathogenic threats, these developments have come at the cost of the breaking of our biosphere, leading to climate catastrophe and social dissolution. It is the kindness, compassion, and forgiveness of animals such as Babe and Esther who are responding to *our* call for aid, who help teach *us* what is required.

This is why being an "advocate for animals" also requires us to be an "animal advocate"—a reminder that we are also vulnerable animals. We need to appeal to nonhuman animals to help us recognize, respect, and stop exploiting each other. *These* advocates are often those who break out of the incomprehensible numbers of farmed animals to remind us, and hearten us, that they are still willing, after everything, to assert their desire to live in "emotional fellowship" with us. That is why Babe—this exploited and abused pig with whom we share so much, who is put to death in the CO_2 killing chambers just as we are gassing ourselves to death by emitting too much greenhouse gas into our thin air—can be an icon of our times.

That said, the future for the pig, for all farmed animals, remains bleak. If their future remains tragic, so ours too will end in tragedy. We urgently need to change our narrative of domination in a fully felt sense. We need to embody a new story. *Change is always bodily change.* And the torture of an immobilized body is also always the torture of that body–mind–world. The question remains for us all: What are we willing to do with our bodies for all of our good, before this thin air we breathe runs out?

Discussion Questions

1. The concept that "all change is bodily change" is at the heart of *The Pig in Thin Air*'s argument for what makes effective activism. How might this idea be interpreted for your own experience of activism?
2. The book presents a number of arguments for identity as the "missing link" in effective animal and environmental activism. What are the ways in which identity processes are currently used in campaigns to inform and influence people's behavior?
3. The author focuses on the processes by which he "woke up" to the horrors of animal agriculture. Does this idea of "waking up" resonate with you particularly? If so, why? This could be in relation to the treatment of animals or any other aspect of culture and society.
4. *The Pig in Thin Air* uses the metaphor of the line to draw attention to the dualistic nature of experience and division between categories, such as man/

woman, human/nonhuman. How is this metaphor helpful in reconceiving the relationships between groups, and what are its limitations?

5. Stories and narrative, and their repetition, are central to how *The Pig in Thin Air* makes its argument for how we come to understand our relations to nonhumans. What are the particular stories that you know or experience that shape your relation with the nonhuman?
6. Emotions are key to how the author came to understand and then share his own story—especially those emotions of embarrassment, shame, pride, and anger. What are the prominent emotions in your experience of activism work, either the work you perform or that you witness?
7. What does bearing witness mean to you? When might bearing witness fail to lead to changes in relations with others?
8. The book seems to suggest that direct encounters with nonhumans are essential for developing our "moral attention" to others, but they cannot take place until we have a "balanced and clear self-concept" of who we are. Does one always need to precede the other? How might the two capacities overlap and interact over time?
9. The author held many expectations and prejudices of what animal activism might be. How have you

experienced your own and others' expectations and prejudices around your form of activism?

10. How could your community engage more directly with learning how "the body keeps the score" of what happens to us in our lives, and what are the ways schools and colleges can play a role in reconnecting individuals with embodied experience and learning through the body?
11. One of the ideas in the book is that vegan food practices are more sustainable for the planet in terms of climate change. How do you or your community currently respond to climate change, and what are the emotions and bodily experiences of these responses?

Notes

1. See <www.peta.com> and <www.peta2.com>.
2. Morell, Virginia. "Causes of the Furred and Feathered Rule the Internet," *National Geographic*, March 14, 2014 <http://news.nationalgeographic.com/news/2014/03/140314-social-media-animal-rights-groups-animal-testing-animal-cognition-world/>.
3. Kotenko, Jam. "From CuteOverload to 'Social Petworks,' Adorable Animals Own the Web," Digital Trends, June 5, 2013 <http://www.digitaltrends.com/social-media/from-cuteoverload-to-social-petworks-adorable-animals-are-taking-over-the-web/#ixzz3gWc1Vs70>.
4. Hauge, Ida Berg. "The Challenge of the Anti-Dairy Movement in Social Media," 2013 <http://www.asuder.org.tr/yayinlar/idf_2013_japan_presentations/pdfler/31_kasim_pazarlama_oturum_3/Ida_Berg_Hauge.pdf>.
5. See <https://vimeo.com/87384748>.
6. Follow the #farm365 stream here: <https://twitter.com/hashtag/farm365>.

7. "Following Along on the #farm365 with Andrew Campbell," *Real Agriculture*, January 2, 2015 <https://www.realagriculture.com/2015/01/following-along-farm365-andrew-campbell/>.
8. Campbell, Andrew. "The Highs & Lows of Week One on #Farm365," *Let's Talk Farm Animals*, January 7, 2015 <http://www.letstalkfarmanimals.ca/2015/01/08/the-highs-lows-of-week-one-on-farm365/>.
9. See <https://www.facebook.com/estherthewonder-pig?fref=ts>.
10. "Determined Pig Jumps from Moving Truck to Escape Slaughterhouse," *Metro*, June 6, 2014 <http://metro.co.uk/2014/06/06/pig-escapes-slaughterhouse-truck-in-guangxi-china-4752625/>.
11. Leo Tolstoy is famous for his sympathies toward animals and is widely quoted from his *Calendar of Wisdom*, New York: Scribner, 1997: "When the suffering of another creature causes you to feel pain, do not submit to the initial desire to flee from the suffering one, but on the contrary, come closer, as close as you can to him who suffers, and try to help."
12. In July 2015, a number of activists from Plane Stupid were arrested after breaking into London's Heathrow Airport to protest the building of a third runway. See "Plane Stupid Activists on Heathrow Runway in Climate Protest," Plane Stupid, July 13, 2015 <http://www.planestupid.com/blogs/2015/07/13/plane-stupid-activists-heathrow-runway-climate-protest>.
13. See "Climate & People," WWF Global <http://wwf.panda.org/about_our_earth/aboutcc/problems/people_at_risk/personal_stories/>.

14. "CRC Launches Climate Witness Project," CRCNA, July 15, 2015 <http://www.crcna.org/news-and-views/crc-launches-climate-witness-project>.
15. See, for example, Ferdman, Roberto A. "Stop Eating So Much Meat, Top U.S. Nutritional Panel Says," *Washington Post*, February 19, 2015.
16. See "Key Facts and Findings," UNFAO <http://www.fao.org/news/story/en/item/197623/icode/>.
17. "Climate Change 2014: Synthesis Report. Contribution of Working Groups I, II and III to the Fifth Assessment Report of the Intergovernmental Panel on Climate Change [Core Writing Team, R.K. Pachauri and L.A. Meyer (eds.)]." IPCC, Geneva, Switzerland, September 2014 <http://www.ipcc.ch/report/ar5/syr/>.
18. "Bald Chicken 'Needs no Plucking,'" *BBC News*, May 21, 2002 <http://news.bbc.co.uk/2/hi/science/nature/2000003.stm>.
19. Renner, Michael. "Peak Meat Production Strains Land and Water Resources," *Vital Signs*, Worldwatch Institute, August 22, 2014 <http://vitalsigns.worldwatch.org/vs-trend/peak-meat-production-strains-land-and-water-resources>.
20. See "New U.S. Clean Power Plan Begs the Question: What About a Clean Food Plan?" Chomping Climate Change, August 3, 2015 <http://www.chompingclimatechange.org/updated-analysis/new-new-u-s-clean-power-plan-begs-the-question-what-about-a-clean-food-plan/>.
21. See Climate Outreach <http://www.climateoutreach.org.uk>.

22. Lawrence, Jeff. "Jane Goodall Urges Vancouver Aquarium to End Cetacean Captivity," CTV News, May 27, 2014 <http://bc.ctvnews.ca/jane-goodall-urges-vancouver-aquarium-to-end-cetacean-captivity-1.1840324>.
23. See *The Last Pig* <http://www.thelastpig.com/>.
24. See Butina CO2 Backloader System <http://www.butina.eu/products/backloader/>.
25. See the Animals Australia investigation at <http://animalsaustralia.org/features/not-so-humane-slaughter/>.
26. See for example, Animal Equality, "International Animal Rights Day 2010 | Madrid (Spain)," August 29, 2011 <https://www.youtube.com/watch?v=cGyrPib-8HI>.
27. See <http://www.thenard.org/>.
28. See <http://www.269life.com>.

Bibliography

Adams, Carol. J. "Activism and the Perils of Burnout: Learning to Take Care of Ourselves," *Satya*, July/August, 2001. Available online at <http://www.satyamag.com/jul01/adams-randour.html>.

———. "The War on Compassion," in *The Feminist Care Tradition in Animal Ethics,* ed. Carol J. Adams and Josephine Donovan. New York: Columbia University Press, 2007, 21–36.

Anderson, Kip, and Keegan Kuhn. *Cowspiracy*. Santa Rosa, Calif.: Aum Films, 2014.

Atwood, Margaret. *Oryx and Crake*. London: Virago, 2003.

Bailly, Jean-Christophe. *The Animal Side.* New York: Fordham University Press, 2011.

Barthes, Roland. *Mythologies.* London: Vintage Classics, 1983.

Baur, Gene. *Farm Sanctuary: Changing Hearts and Minds About Animals and Food.* New York: Touchstone Books, 2008.

Best, Steven, and A. John Nocella. *Terrorists or Freedom Fighters? Reflections on the Liberation of Animals.* New York: Lantern Books, 2004.

Birke, Lynda. "Naming Names—or, What's In It for the Animals?" *Humanimalia* 1(1) (2009): 1–9.

Blackman, Lisa. *The Body: Key Concepts.* Oxford and New York: Berg, 2008.

———. *Immaterial Bodies: Affect, Embodiment, Mediation.* London: Sage, 2012.

Blustein, Jeffrey. *The Moral Demands of Memory.* New York: Cambridge University Press, 2008.

Booth, W. James. *Communities of Memory: On Witness, Identity, and Justice.* Ithaca, N.Y.: Cornell University Press, 2006.

Bradshaw, Gay. "Turkey Like Me," in *Turning Points in Compassion: Personal Journeys of Animal Advocates,* ed. Gypsy Wulff and Fran Chambers. Perth, Australia: Spirit Wings Publishing, 2015, 52–56.

Brown, Michael, ed. *Little Book of Curiosity: Game of Phones.* London: Universal McCann, 2015.

Buller, Henry. "Individuation, the Mass and Farm Animals," *Theory, Culture and Society* 30(7–8) (2013): 155–175.

Burt, Jonathan. "Conflicts and Slaughter in Modernity," in *Killing Animals,* ed. The Animal Studies Group. Champaign: University of Illinois Press, 2006, 120–144.

Butler, Judith. *Gender Trouble: Feminism and the Subversion of Identity*. London and New York: Routledge, 2006.

Calarco, Matthew. *Thinking Through Animals: Identity, Difference, Indistinction.* Redwood, Calif.: Stanford University Press, 2015.

Coe, Sue. *Union*. Graphite, gouache, and watercolor on paper. New York: Galerie St. Etienne, 1989.

Coetzee, J. M. *Disgrace*. London: Vintage, 2000.

———. *The Lives of Animals*. London: Profile Books, 2000.

Cole, Matthew, and Kate Stewart. *Our Children and Other Animals: The Cultural Construction of Human–Animal Relations in Childhood.* Farnham: Ashgate, 2014.

Cooney, Nick. *Change of Heart: What Psychology Can Teach Us About Spreading Social Change.* New York: Lantern Books, 2011.

Cudworth, Erika. *Social Lives with Other Animals: Tales of Sex, Death and Love.* London: Palgrave Macmillan, 2011.

Cvetkovich, Ann. *Depression: A Public Feeling.* Durham and London: Duke University Press, 2012.

Darrieussecq, Marie. *Pig Tales.* London: Faber and Faber, 2003.

Daston, Lorraine, and Gregg Mitman. *Thinking with Animals: New Perspectives on Anthropomorphism.* New York: Columbia University Press, 2006.

Dean, Jodi. *Blog Theory: Feedback and Capture in the Circuits of Drive.* Cambridge, U.K.: Polity Books, 2010.

Derrida, Jacques. *The Animal That Therefore I Am.* New York: Fordham University Press, 2008.

Despret, Vinciane, and Jocelyne Porcher. *Être bête.* Arles: Actes Sud, 2007.

Diamond, Cora. "The Difficulty of Reality and the Difficulty of Philosophy," in *Philosophy & Animal Life*, ed. Cary Wolfe. New York: Columbia University Press, 2008, 43–90.

———. "Eating Meat and Eating People," *Philosophy* 53 (1978): 465–479.

Donaldson, Sue, and Will Kymlicka. *Zoopolis: A Political Theory of Animal Rights.* Oxford: Oxford University Press, 2011.

Donovan, Josephine. "Sympathy and Interspecies Care: Toward a Unified Theory of Eco- and Animal

Liberation," in *Critical Theory and Animal Liberation*, ed. John Sanbonmatsu. Plymouth, U.K.: Rowman & Littlefield, 2011, 277–294.

Donovan, Josephine, and Carol J. Adams, eds. *The Feminist Care Tradition in Animal Ethics*. New York: Columbia University Press, 2007.

Essig, Mark. *Lesser Beasts: A Snout-to-Tail History of the Humble Pig*. New York: Basic Books, 2015.

Faber, Michel. *Under the Skin*. Edinburgh: Canongate Books, 2014.

Felman, Shoshana, and Dori Laub. *Testimony*. London and New York: Routledge, 1992.

Grandin, Temple, ed. *Improving Animal Welfare: A Practical Approach, Second Edition*. Boston: CAB International, 2015.

Haraway, Donna. *The Companion Species Manifesto: Dogs, People, and Significant Otherness*. Chicago: University of Chicago Press, 2003.

Hawthorne, Mark. *Bleating Hearts: The Hidden World of Animal Suffering*. Winchester, U.K.: Changemakers Books, 2013.

———. "How Do Graphic Images Affect Animal Advocacy?" 2012. Available online at <https://strikingattheroots.wordpress.com/2012/11/01/how-do-graphic-images-affect-animal-advocacy/>.

———. *Striking at the Roots: A Practical Guide to Animal Activism*. Winchester, U.K.: Zero Books, 2008.

Hirst, Damien. *This Little Piggy Went to Market, This Little Piggy Stayed at Home*. Glass, pig, painted steel, silicone, acrylic, plastic cable ties, stainless steel, formaldehyde solution, and motorized painted steel base. London: Saatchi Gallery, 1996.

Höijer, Birgitta. "Emotional Anchoring and Objectification in the Media Reporting on Climate Change," *Public Understanding of Science* 19(6) (2009): 717–731.

Goodland, Robert, and Jeff Anhang. *Livestock and Climate Change: What If the Key Actors in Climate Change Are . . . Cows, Pigs, and Chickens?* Washington, D.C.: World Watch, 2009.

Graham-Leigh, Elaine. *A Diet of Austerity: Class, Food and Climate Change.* Winchester, U.K.: Zero Books, 2015.

Gruen, Lori. *Entangled Empathy.* New York: Lantern Books, 2015.

Ingold, Tim. *Being Alive: Essays on Movement, Knowledge and Description.* London: Routledge, 2011.

Jaspal, Rusi, Brigitte Nerlich, and Marco Cinnirella. "Human Responses to Climate Change: Social Representation, Identity and Socio-Psychological Action," *Environmental Communication* 8(1) (2014): 110–130.

Jenni, Kathie. Commentary on "The Case for Animal Rights," 2011. Available online at <http://nationalhumanitiescenter.org/on-the-human/2011/05/regan-preface/>.

jones, pattrice. *The Oxen at the Intersection: A Collision.* New York: Lantern Books, 2014.

Joy, Melanie. "Discovering Carnism," in *Turning Points in Compassion: Personal Journeys of Animal Advocates,* ed. Gypsy Wulff and Fran Chambers. Perth, Australia: Spirit Wings Publishing, 2015, 141–145.

———. *Why We Love Dogs, Eat Pigs, and Wear Cows.* San

Francisco: Conari Press, 2010.

Kahan, Dan. "Cultural Cognition as a Conception of the Cultural Theory of Risk," in *Handbook of Risk Theory: Epistemology, Decision Theory, Ethics and Social Implications of Risk*, ed. Sabine Roeser, Rafaela Hillerbrand, Per Sandin, and Martin Peterson. London: Springer, 2012, 725–760.

Kashdan, Todd. *Curious? Discover the Missing Ingredient to a Fulfilling Life.* New York: William Morrow, 2009.

Kemmerer, Lisa. *Eating Earth: Environmental Ethics and Dietary Choice.* Oxford: Oxford University Press, 2015.

Klein, Naomi. *This Changes Everything: Capitalism vs. the Climate.* New York: Simon & Schuster, 2014.

Lakoff, George, and Mark Johnson. *Philosophy in the Flesh: The Embodied Mind and Its Challenge to Western Thought.* New York: Basic Books, 1999.

Lamont, Robin. *The Chain.* New York: Grayling Press, 2014.

Lappé, Anna. *Diet for a Hot Planet: The Climate Crisis at the End of Your Fork and What You Can Do About It.* New York: Bloomsbury, 2006.

Lappé, Frances Moore. *Diet for a Small Planet.* New York: Ballantine Books, 1971.

Luke, Brian. "Justice, Caring, and Animal Liberation," *Between the Species* 8(2) (1992): 100–109.

Lymbery, Philip, and Isabel Oakeshott. *Farmageddon: The True Cost of Cheap Meat.* London: Bloomsbury, 2014.

Malcolmson, Robert, and Stephanos Mastoris. *The English Pig: A History.* London: Hambledon and London, 2001.

Marshall, Liz. *The Ghosts in Our Machine.* Toronto: LizMars

Productions, 2014.

McArthur, Jo-Anne. *We Animals.* New York: Lantern Books, 2013.

McGlone, John J. "The Future of Pork Production in the World: Toward Sustainable, Welfare-Positive Systems," *Animals* 3 (2013): 401–415.

McHugh, Susan. *Animal Stories.* Minneapolis: University of Minnesota Press, 2011.

McQueen, Alison. "Compassion and Tragedy in the Aspiring Society," *Phenomenology and the Cognitive Sciences* 13 (2014): 651–657.

Midgley, Mary. "Persons and Non-Persons," in *In Defense of Animals*, ed. Peter Singer. New York: Basil Blackwell, 1985, 52–62.

Mitchell, David. *Cloud Atlas*. London: Random House, 2004.

Mizelle, Brett. *Pig.* London: Reaktion Books, 2011.

Moser, Susanne C., and Lisa Dilling. "Communicating Climate Change: Closing the Science–Action Gap," in *The Oxford Handbook of Climate Change and Society*, ed. Richard B. Norgaard, David Schlosberg, and John S. Dryzek. New York: Oxford University Press, 2011.

Murakami, Haruki. *What I Talk About When I Talk About Running*. London: Vintage Books, 2009.

Murdoch, Iris. *Metaphysics as a Guide to Morals*. New York: Viking Penguin, 1993.

Noske, Barbara. *Humans and Other Animals: Beyond the Boundaries of Anthropology.* London: Pluto Press, 1989.

Nussbaum, Martha. *Political Emotions: Why Love Matters for Justice*. Cambridge, Mass.: Harvard University Press, 2013.

O'Neill, Saffron, and Sophie Nicholson-Cole. "Fear Won't Do It: Promoting Positive Engagement with Climate Change Through Visual and Iconic Representations," *Science Communication* 30(3) (2009): 335–379.

Oppenlander, Richard A. *Comfortably Unaware: Global Depletion and Food Responsibility . . . What You Choose to Eat Is Killing Our Planet.* Minneapolis: Langdon Street, 2011.

Orwell, George. *Animal Farm.* London: Penguin Classics, 2000 [1945].

Pachirat, Timothy. *Every Twelve Seconds: Industrialized Slaughter and the Politics of Sight.* New Haven and London: Yale University Press, 2011.

Patrick-Goudreau, Colleen. *On Being Vegan: Reflections on a Compassionate Life.* Oakland, Calif.: Montali Press, 2013.

Pick, Anat. *Creaturely Poetics: Animality and Vulnerability in Literature and Film.* New York: Columbia University Press, 2011.

Plumwood, Valerie. "Human Vulnerability and the Experience of Being Prey," *Quadrant*, March 1995, 29–34.

Plutarch. *Gryllus.* Available online at <http://penelope.uchicago.edu/Thayer/E/Roman/Texts/Plutarch/Moralia/Gryllus*.html>.

Probyn, Elspeth. "The Cultural Politics of Fish and Humans: A More-Than-Human Habitus of Consumption," *Cultural Politics* 10(3) (2014): 287–299.

Reynolds, Mark. *Running with the Pack: Thoughts from the Road on Meaning and Mortality.* London: Granta, 2013.

Rowe, Martin, ed. *Running, Eating, Thinking: A Vegan Anthology.* New York: Lantern Books, 2014.

Ruddick, Sara. "Maternal Thinking," *Feminist Studies* 6(2) (1980): 342–367.

Sanbonmatsu, John, ed. *Critical Theory and Animal Liberation*. Plymouth, U.K.: Rowman & Littlefield, 2011.

Shilling, Chris. *The Body and Social Theory*. London: Sage, 2012.

Shukin, Nicole. *Animal Capital: Rendering Life in Biopolitical Times*. Minneapolis: University of Minnesota Press, 2009.

Simon, David Robinson. *Meatonomics.* San Francisco: Conari Press, 2013.

Sinclair, Upton. *The Jungle*. Stilwell, Ks.: Digireads Books, 2007 [1906].

Sontag, Susan. *On Photography*. New York: Farrar, Straus and Giroux, 1977.

Springgay, Stephanie. "The Ethico-Aesthetics of Affect and a Sensational Pedagogy," *Journal of the Canadian Association for Curriculum Studies* 9(1) (2011): 66–82.

Stallybrass, Peter, and Allon White. *Poetics and Politics of Transgression*. London: Routledge, 1986.

Stone, Gene, and Jon Doyle. *The Awareness*. New York: The Stone Press, 2014.

Tessman, Lisa. *Burdened Virtues: Virtue Ethics for Liberatory Struggles*. New York: Oxford University Press, 2005.

Thieme, Marianne. *Meat the Truth*. Amsterdam: Nicolaas G. Pierson Foundation, 2007.

Thoreau, Henry David. *Walden*. Oxford: Oxford Paperbacks, 2008 [1854].

UNFAO. *Livestock's Long Shadow: Environmental Issues and Options.* Rome: UNFAO, 2006.

van der Kolk, Bessel. *The Body Keeps the Score: Brain, Mind, and Body in the Healing of Trauma*. New York: Viking Books, 2014.

Walker, Margaret Urban. "Moral Understandings: Alternative 'Epistemology' for a Feminist Ethics," in *Explorations in Feminist Ethics,* ed. Eve Browning Cole and Susan Coultrap-McQuin. Bloomington: Indiana University Press, 1992, 165–175.

Whitlock, Gillian. "Protection," in *We Shall Bear Witness: Life Narratives and Human Rights,* ed. Meg Jensen and Margaretta Jolly. Madison: University of Wisconsin Press, 2014, 80–99.

Wilkie, Rhoda. "Multispecies Scholarship and Encounters: Changing Assumptions at the Human–Animal Nexus," *Sociology* 49(2) (2015): 323–329.

Winnicott, Donald Woods. "Ego Distortion in Terms of True and False Self," in *The Maturational Process and the Facilitating Environment: Studies in the Theory of Emotional Development*. New York: International Universities Press, 1965, 140–157.

Woolf, Virginia. *Moments of Being*. London: Triad Grafton Books, 1978.

Wordsworth, William. "Ode: Intimations of Immortality," in *Poems: in Two Volumes*, 1804.

Wright, Kate. "Becoming-With," *Environmental Humanities* 5 (2014): 277–281.

Wulff, Gypsy, and Fran Chambers, eds. *Turning Points in Compassion: Personal Journeys of Animal Advocates*. Perth, Australia: Spirit Wings Publishing, 2015.

About the Author

Alex Lockwood, Ph.D., is a writer, educator, and activist working in the fields of literature, creative writing, media, environment, and Human–Animal Studies. He has published widely on Rachel Carson and her legacy for contemporary writers. He lives in Newcastle, United Kingdom, where he writes about animals, vegan life practices, and running for the *Guardian*, *Like the Wind* magazine, *Earthlines,* and other publications. He is a director of the Vegan Lifestyle Association and a member of the Research Advisory Committee for the Vegan Society.

About the Publisher

LANTERN BOOKS was founded in 1999 on the principle of living with a greater depth and commitment to the preservation of the natural world. In addition to publishing books on animal advocacy, vegetarianism, religion, and environmentalism, Lantern is dedicated to printing books in the U.S. on recycled paper and saving resources in day-to-day operations. Lantern is honored to be a recipient of the highest standard in environmentally responsible publishing from the Green Press Initiative.

lanternbooks.com

Lightning Source UK Ltd.
Milton Keynes UK
UKOW04f2207070316

269776UK00001B/2/P